Just-In-Time Accounting

JUST-IN-TIME ACCOUNTING

How to Decrease Costs and Increase Efficiency

Third Edition

STEVEN M. BRAGG

WILEY

John Wiley & Sons, Inc.

Published by John Wiley & Sons, Inc., Hoboken, New Jersey.

Published simultaneously in Canada.

For general information on our other products and services, or technical support, please contact our Customer Care Department within the United States at 800-762-2974, outside the United States at 317-572-3993 or fax 317-572-4002.

Wiley also publishes its books in a variety of electronic formats. Some content that appears in print may not be available in electronic books.

For more information about Wiley products, visit our Web site at http://www.wiley.com.

Library of Congress Cataloging-in-Publication Data:

Bragg, Steven M.

Just-in-time accounting: how to decrease costs and increase efficiency/Steven M. Bragg. – 3rd ed.

p.   cm.

Includes index.

ISBN 978-0-470-40372-3 (cloth)

1. Just-in-time systems–Accounting.   I. Title.

HF5686.M3B68 2009

657–dc22

2008042935

Printed in the United States of America

10 9 8 7 6 5 4 3 2 1

To my Lifelong Partner, Melissa

Contents

About the Author

Steven Bragg, CPA, has been the chief financial officer or controller of four companies, as well as a consulting manager at Ernst & Young and auditor at Deloitte & Touche. He received a Master's degree in Finance from Bentley College, an MBA from Babson College, and a Bachelor's degree in Economics from the University of Maine. He has been the two-time president of the Colorado Mountain Club, and is an avid alpine skier, mountain biker, and certified master diver. Mr. Bragg resides in Centennial, Colorado. He has written the following books through John Wiley & Sons:

Accounting and Finance for Your Small Business

Accounting Best Practices

Accounting Control Best Practices

Accounting for Payroll

Accounting Reference Desktop

Billing and Collections Best Practices

Business Ratios and Formulas

Controller's Guide to Costing

Controller's Guide to Planning and Controlling Operations

Controller's Guide: Roles and Responsibilities for the New Controller

Controllership

Cost Accounting

Design and Maintenance of Accounting Manuals

Essentials of Payroll

Fast Close

Financial Analysis

GAAP Guide

GAAP Implementation Guide

Inventory Accounting

Inventory Best Practices

Just-in-Time Accounting

Management Accounting Best Practices

Managing Explosive Corporate Growth

Mergers and Acquisitions

Outsourcing

Payroll Best Practices

Revenue Recognition

Sales and Operations for Your Small Business

The Controller's Function

The New CFO Financial Leadership Manual

The Ultimate Accountants' Reference

Throughput Accounting

Also:

Advanced Accounting Systems (Institute of Internal Auditors)

Investor Relations (Accounting Tools)

Run the Rockies (CMC Press)

Preface

This third edition of *Just-in-Time Accounting* is designed for anyone who wants to streamline an accounting system so that transactions can be processed with minimal errors and staff time. The book uses a multi-step approach to improving accounting systems. First, we describe and chart an existing process, focusing on the inputs, processing steps, and outputs associated with each transaction. In many cases, this includes an analysis of wait times and paper flow among employees and departments, in order to differentiate non value-added from value-added activities. Second, we list a number of suggestions for reducing or streamlining the workload. This can include the application of new technology, eliminating redundant or unnecessary control points, reducing the number of people involved, and compressing some activities.

Each chapter outlines the effects on accounting controls of the suggestions for improvement. This includes a discussion of which controls can be eliminated, how to bolster remaining controls, and which new controls should be added.

Each chapter contains samples of cost/benefit analyses that can be used as models when creating actual cost/benefit analyses for justifying the implementation of revised systems. Special problems with identifying some costs and associated savings are also noted.

Streamlining the accounting function eliminates some of the old accounting reports. In many cases, this requires their replacement with new reports that are more useful for the revised system. Consequently, sample reports are described in each chapter, which can be used as templates. In addition, the chapters include new metrics that monitor the revised systems.

This multi-step approach applies to Chapters 2 through 8, covering transactions in the following areas: the sales cycle, cash, inventory, accounts payable, cost accounting, payroll, and the budget. Chapters 9 through 12 discuss improvements to other accounting areas that either require streamlining or are needed in the streamlining effort: closing the books, data collection and storage systems, process documentation, and change management.

The introductory chapter contains an overview of how to properly organize a new accounting department to achieve maximum levels of efficiency. The activities addressed include the creation of job descriptions, performance reviews, work calendars, policies and procedures, training programs, and work flow analysis—all basic "blocking and tackling" issues that are all too frequently ignored.

This book is intended for accounting managers who want to improve the performance of the accounting department. It describes how to integrate best practices into accounting systems, resulting in remarkable gains in operating efficiencies.

STEVEN M. BRAGG
Centennial, Colorado
November 2008

Free Online Resources

Steve offers a broad array of free accounting resources at www. accountingtools.com. The site includes dozens of *Accounting Best Practices* podcast episodes, as well as hundreds of best practices articles. It also contains control charts, process flows, costing methodologies, job descriptions, metrics, and much more.

Chapter 1

Setting Up and Improving the Accounting Department

Before we can dig into the minutia of how to improve each of the various accounting functions, it is first necessary to set up the infrastructure to ensure that the accounting department will function in a reliable manner. These basic tasks are ignored all too frequently when first setting up an accounting department.

In many instances, accountants are given a pile of work to complete, with no hint as to the priority of each item or how they interrelate. The result is a daily scramble to complete whatever is "hot" on that day. Assignments of accounting staff members are often changed with excessive regularity, so that no one acquires an area of expertise in which he or she can improve skills and overall level of confidence. As a direct result of this unstructured work pattern, the number of transactional errors is extremely high; employees do not have the expertise to "lock in" the procedures and related training that will keep errors from occurring. Once errors are found, considerable staff time must be reallocated to fixing them. Thus, we enter into a vicious circle of constant daily problems, with the department always struggling to keep up with the workload. The simple organizational techniques outlined in this chapter can resolve many of these problems.

FORECAST CASH

The single most important priority is forecasting the flow of cash. Without money to pay employees or suppliers, a company will very quickly find itself out of business. All other considerations are secondary to this issue. Accordingly, there should be a weekly cash forecast, such as the one shown in Exhibit 1.1. This report reveals the time periods when there may be cash flow difficulties, thereby enabling management to plan well in advance for additional financing or other activities that will avoid any cash shortfalls. This format can vary considerably by type of business, and may be an automated report in some accounting systems.

The forecast in Exhibit 1.1 starts with a sales forecast for each company subsidiary, showing expectations for the next eight weeks. Next is a breakdown of the largest accounts receivable items by expected date of cash receipt, as well as a single "Cash, Minor Invoices" line item that summarizes all other small cash receipts. The next section lists all expected cash outflows from payroll, payables, and capital expenditures. The bottom of the spreadsheet lists the rolling cash balance at the end of each reporting period.

This forecast should be verified with the sales staff, collections personnel, and purchasing employees to ensure that the numbers are accurate. Accounts payable tend to be the most accurate, since these figures are under the direct control of management, and can be managed to some extent to fit the requirements of the forecast. Also, compare previous cash forecasts to actual results to see what differences arose. By incorporating these differences into future reports, the cash forecast can gradually become a reasonably accurate indicator of future results.

The report is hand-delivered to key recipients each week, along with a verbal or written discussion of any key problem areas. Leaving

Exhibit 1.1 Sample Cash Forecast

				Company Cash Forecast				
				For the Week Beginning on				
Week Beginning Date	9/6/2009	9/13/2009	9/20/2009	9/27/2009	10/4/2009	10/11/2009	10/18/2009	10/25/2009
Beginning Cash	$ 300,000	$ 142,992	$ 32,022	$ 171,485	$ 6,077	$ 3,660	$ 698,445	$ 656,380
Receipts from Sales Projections:								
ABC Subsidiary						$ 170,500		
DEF Subsidiary						$ 500,000		
GHI Subsidiary						$ 584,425		
Uncollected Invoices:								
Amber Exploration		$ 63,667		$ 62,501		$ 64,975		
Botany Bay Drilling		$ 38,425		$ 18,872	$ 12,521			
Callow Consulting Co.			$ 100,472					
Deep Drilling Divers		$ 41,290						
Eastern Europe Pipeline	$ 62,976	$ 53,135	$ 24,772	$ 6,676	$ 22,327	$ 33,816		
Franklin Moss Consulting	$ 81,005		$ 20,440		$ 29,500		$ 14,935	
Guttering Oil and Son		$ 54,564		$ 55,000				
Hinter Drilling		$ 80,250			$ 21,204			
Indian Express Pipe Repair		$ 121,360			$ 99,231			
Cash, Minor Invoices	$ 5,380	$ 14,029	$ 28,990	$ 48,044				
Total Cash In	$ 178,784	$ 466,720	$ 174,674	$ 191,093	$ 184,783	$ 1,353,716	$ 14,935	$ —

(Continued)

Exhibit 1.1   Continued

Company Cash Forecast

For the Week Beginning on

Cash Out:								
Payroll + Taxes		$ 330,500		$ 331,500		$ 332,500		$ 332,500
401k Payments		$ 32,000			$ 32,000	$ 32,000		
Commissions		$ 12,000						
Contractors	$ 26,628	$ 168,190				$ 279,431		
Rent					$ 53,000			
Medical Insurance	$ 69,200				$ 69,200			
Capital Purchases	$ 81,500				$ 8,000			
Expense Reports	$ 5,000	$ 15,000	$ 5,000	$ 5,000	$ 5,000	$ 15,000		$ 5,000
Other Expenses	$ 153,464	$ 20,000	$ 30,211	$ 20,000	$ 20,000			$ 20,000
Total Cash Out:	$ 335,792	$ 577,690	$ 35,211	$ 356,500	$ 187,200	$ 658,931		$ 357,500
Net Change in Cash	$ (157,008)	$ (110,970)	$ 139,463	$ (165,408)	$ (2,417)	$ 694,785	$ (42,065)	$ (357,500)
Ending Cash:	$ 142,992	$ 32,022	$ 171,485	$ 6,077	$ 3,660	$ 698,445	$ 656,380	$ 298,880

the report on someone's desk is an invitation for future trouble, since the recipient may not see the prime indicators of a cash shortfall that are so evident to the person who prepared it.

REVIEW CONTRACTS

A complete examination of all unexpired legal agreements ranks high on the accounting priority list. These may contain any number of requirements for cash expenditures or receipts that should be included in the cash forecast, such as a balloon payment on a debt agreement, or a scheduled increase in a rent payment. Another example is a periodic minimum royalty payment. If not carefully tracked, these issues can result in very unpleasant and sudden crises, as well as missed opportunities to collect additional funds.

Rather than just reviewing legal agreements in a cursory manner, it is better to organize them into a summary–level format, such as the one shown in Exhibit 1.2. This summary should be included in the standard calendar of department activities (described later), so that actions can be scheduled based on the agreements. By using the exhibit, a manager can quickly determine the dates when payments are due, when key contractual dates occur, and the names of the other parties to the agreements. These are the key factors to be aware of when incorporating legal agreements into the accounting department's activities. Organize the filing system for the legal documents that are referenced in the summary, so that anyone can quickly access the original document.

Better yet, scan all legal agreements and store the scanned documents in a file server, where they are easily accessible from remote locations. Then store the original documents in a locked fireproof safe, so that no one can remove or misfile them.

Proper custody and knowledge of the contents of legal agreements is a fundamental requirement for the proper management of the accounting department.

Exhibit 1.2   Sample Format for Legal Summary

Party Name	Start/Stop Dates	Revenues or Expenditures	Description
Acme Leasing	Jan. 2009 – Dec. 2012	$504.52/mo, due on 15th of the month	Copier lease. Requires balloon payment of $5,400 on 12/31/2012
GEM Leasing	Ongoing	$228.15/mo, due on 10th of the month	Storage trailer lease. Ongoing. Review need for trailer each quarter.
Bartony Design Co.	June 1, 2010 Renewal	8% of sales on Kid-Jump product, due on 20th of the month	Royalty on Kid-Jump product, payable as long as there are product sales. Agreement renews on 6/01/2011.
Play & Go Inc.	Lifetime franchise	4% of sales, incoming franchise fee, due on 5th of the month	Franchise fee paid by operator of Play & Go store, based on percentage of net sales.

ESTABLISH JOB DESCRIPTIONS

Accounting staffs do not always know precisely what they are supposed to do. Instead, they are hired, trained briefly in a few key tasks,
and then left alone. They do not know if there are additional tasks to
be completed on a less frequent basis, nor do they know when tasks
are to be completed, or what constitutes a good job. This has several
ramifications. First, it is impossible to review the work of such a person when there is no baseline against which to judge. Second, key
tasks that are performed infrequently tend to be completely ignored,
because no responsibility has been assigned. Third, it is difficult to
determine an employee's pay scale when there is uncertainty about
the boundaries of that person's job. Finally, many employees do not
know for certain to whom they report. All of these factors present a

---

Exhibit 1.3   Sample Job Description

Position Name: Cost Accountant

Reports to: Assistant Controller

Supervises: None

Basic Function: This position is accountable for the ongoing analysis of process constraints, target costing projects, and margin analysis. The cost accountant must also construct and monitor those cost-effective data accumulation systems needed to provide an appropriate level of costing information to management.

Principal Accountabilities:

1. Construct data accumulation systems for a cost accounting system
2. Create and review the controls needed for data accumulation and reporting systems
3. Conduct ongoing process constraint analyses
4. Update bill of material standard costs
5. Report on breakeven points by products, work centers, and factories
6. Report on margins by product and division
7. Report on periodic variances and their causes
8. Analyze capital budgeting requests
9. Perform cost accumulation tasks as a member of the target costing group
10. Coordinate physical inventories and cycle counts
11. Accumulate and apply overhead costs as required by generally accepted accounting principles

Update Date: July 1, 2009

---

strong case in favor of creating a job description for every employee as soon as possible.

An example of a job description is shown in Exhibit 1.3. It states reporting relationships, gives a general overview of the position, and then itemizes specific accountabilities. It is also useful to list the last date on which the job description was updated, which can be used as a trigger to periodically review the description.

ISSUE PERFORMANCE REVIEWS

The job description does not provide a sufficient amount of detail about exactly what level of performance is expected to conduct a performance review. To use the job description in Exhibit 1.3 as an example, item number one is "construct data accumulation systems." Does this mean that all possible systems must be created to give the cost accountant a good review, or is some lesser achievement acceptable? How will this translate into a salary adjustment? If neither the supervisor nor the employee knows the answers to these questions, the expectations of both parties may be far apart, resulting in an interesting performance review.

To resolve this problem, it is necessary to precisely define expected performance levels in advance. This information should be developed as soon as the job description is complete, and then be reviewed with the employee in detail to make sure that there is no confusion about the expectations for each item. The description of these performance levels takes considerable effort to develop; the effort is needed, since they are the basis for a great deal of ensuing employee activity that is vital to the operation of the department.

One way to write performance level expectations is to list them in terms of low, medium, and exceptional performance. Each of the categories should be thoroughly described. Exhibit 1.4 contains a review that is based on item 1 in Exhibit 1.3—for the cost accountant to create data collection systems.

The cost accountant's performance review criteria will take a great deal of time to complete if all other job functions are described in the same manner as the example. Nonetheless, this level of detail is needed to convey to the employee the exact expectations for the job. Since job requirements will change, the manager must review and adjust these written expectations frequently. Each time a change is made, the manager must review the changes with employees.

The performance review format can also be expanded to include the precise amount of pay change that can be expected as a result of meeting each objective. For example, the low performance category in

Exhibit 1.4 Example of Performance Level Expectations

Performance Expectations:
1. Create costing data collection systems.

   a. Low performance. The cost accountant creates no additional data collection systems beyond those already in place, but may create plans for new ones.

   b. Average performance. The cost accountant creates new data collection systems, but the level of data accuracy is not increased by more than 10% or the amount of labor required to collect the data does not decrease by more than 20%.

   c. Exceptional performance. The cost accountant creates new data collection systems that result in data accuracy improvement of more than 10% and reduction in data collection labor of more than 20%.

Exhibit 1.4 might have listed alongside it a pay decrease of 1%, with average performance worthy of no change, and a 2% pay increase associated with the exceptional performance level. Using this approach, an employee can tell precisely how much of a pay change will be occurring before a pay review even begins, which takes all of the tension and uncertainty out of the process. However, this approach can result in excessively high or low pay changes if the person preparing the document does not pay sufficient attention to the difficulty of completing those tasks against which automatic pay changes are listed.

CREATE WORK CALENDARS

Even with a job description, an employee does not know *when* tasks are to be completed. It may be necessary to always complete a report on a Monday, so that it is available for a Tuesday meeting, or perhaps one employee has to complete a deliverable so that it can be used as input to some other process by a different employee on the following day. These are issues that have a major impact on the efficiency of the entire department.

A good way to handle the timing of work steps is to schedule them on an individual work calendar that is handed out to each employee. There should be two calendars—one that lists the major tasks to be completed on a monthly basis, and another that itemizes the daily tasks within each month. An example of a monthly calendar is shown in Exhibit 1.5 . This calendar itemizes those tasks that are not necessarily repeated constantly, such as government reports that need only be created once a year (such as the VETS-100 report in August that lists the number of employees who are veterans), and which could easily be forgotten if not itemized.

A monthly calendar of activities contains all of the tasks noted on the annual calendar, plus all of the continuing daily activities that are

Exhibit 1.5   Annual Calendar of Activities

January	February	March	April
1st Commissions	1st Commissions	1st Commissions	1st Commissions
10th 1099 Forms	20th Merker royalty	15th 401(k) Enrollment	15th Property
15th NM Sales Tax	28th Property Taxes	22nd Audit Begins	20th Merker Royalty
20th UT Sales Tax	28th Trademark review		
May	**June**	**July**	**August**
1st Commissions	1st Commissions	1st Commissions	1st Commissions
20th Rent increase	15th 401(k) Enrollment	15th NM Sales Tax	10th VETS-100 Report
31st Update Procedures	28th Property Taxes	20th UT Sales Tax	20th Merker Royalty
September	**October**	**November**	**December**
1st Commissions	1st Commissions	1st Commissions	1st Commissions
15th 401(k) Enrollment	10th Initial Budget Mtg	30th Budget Due	15th 401(k) Enrollment
20th Merker Royalty		30th Update Procedures	30th Mail W-9 Forms

repeated within the department. An example is shown in Exhibit 1.6. Of the two calendars, this one is used much more frequently; it becomes a daily reference for every employee.

Though the concept of the calendar is an extremely simple one, it is highly effective, for it ensures that the accounting staff completes its assigned tasks on time, every time. However, this degree of effectiveness will continue only if the accounting manager constantly updates these calendars. This typically means that a new calendar should be issued to all accounting employees at the end of every month or quarter.

Exhibit 1.6 Monthly Calendar of Activities

Monday	Tuesday	Wednesday	Thursday	Friday
2	**3**	**4**	**5**	**6**
Cash Forecast	Issue Financials	Print AP Checks	Job Profit Report	Open AR Review
Flash Report	Staff Meeting	Pay Sales Taxes		
Pay Rent				
9	**10**	**11**	**12**	**13**
Cash Forecast	Staff Meeting	Print AP Checks	GSA Payment	Open AR Review
Flash Report		Process Payroll		Update GSA Schedule
				Issue Paychecks
16	**17**	**18**	**19**	**20**
Cash Forecast	Staff Meeting	Print AP Checks	Holiday	Open AR Review
Flash Report				
23	**24**	**25**	**26**	**27**
Cash Forecast	Staff Meeting	Print AP Checks	Pay Summary Report	Open AR Review
Flash Report		Process Payroll		Issue Paychecks

CREATE POLICIES AND PROCEDURES

Job descriptions and work calendars tell an employee what to do and when to do it, but they do not contain any details regarding *how* to perform any tasks. This is not a problem for an employee who has been working in the department for a long time. However, new employees or those who are filling in on unfamiliar tasks on a short-term basis need assistance.

One possibility is to assign an experienced employee the task of providing training to anyone who needs it, but this is a very expensive proposition, and will not work well if there are too few employees who possess a comprehensive knowledge of procedures. A better approach is to document all accounting policies and procedures in a manual that is issued to all members of the department. This codifies all information needed to complete the daily tasks of the department. It can be posted online for ready access by employees, where its text can also be more readily updated.

This manual takes considerable time to complete, since each person in the department must be carefully interviewed regarding his or her work, which must then be translated into both text and a flowchart that adequately explains what he or she does. Also, this information requires constant updating, since procedures change over time. Despite the time required for these activities, it is highly profitable to have up-to-date policy and procedure manuals available to the accounting staff at all times. Policies and procedures are addressed further in Chapter 11, Process Documentation, and much more comprehensively in the author's *Accounting Policies and Procedures Manual.*

ESTABLISH TRAINING SCHEDULES

All of the groundwork for running an effective accounting department has now been laid—job descriptions, performance criteria, work calendars and a policies and procedures manual. However, employees may lack the skills required to use these tools to the fullest degree. There are two types of training that will overcome these issues:

1. *Orientation training* for new employees or those who are transferring to a new position. This relatively short seminar presents each person with the various tools already described in this chapter. Conducted one-on-one, this training ensures that an employee gets up-to-speed as soon as possible. There should be a series of one-on-one follow-up training sessions for the first few weeks of each person's new job, so that any questions regarding unfamiliar tasks can be adequately answered. A vital part of these ongoing sessions is the development of a training plan by the trainer that identifies the weaknesses in each person's job knowledge; this training plan becomes the foundation for the second type of training, which is skills enhancement.

2. *Skills enhancement training* is specifically designed to eliminate any weaknesses in each employee's package of skills, rather than being a broad-based set of training requirements that all employees must complete. For example, rather than requiring all employees to attend a seminar on the legal requirements of reporting foreign exchange transactions (which is usually of interest to only a select few), some employees may attend a remedial English class, while others will need to improve their electronic spreadsheet skills. These are much more basic issues than the high-level topics that are typically taught to employees.

Determining training requirements at this much more basic and individualized level requires very specific training programs that are tailored to the needs of each employee. Here are some options for locating those training programs:

- *College degree classes.* A small number of carefully-selected classes works best; paying for an entire program of study is expensive, and probably imparts more knowledge than is needed for a specific job.
- *Continuing professional education (CPE).* CPE is a relatively inexpensive option that is targeted at very narrow topics, and which can involve seminars or home study. A number of CPE providers are listed at www.accountingtools.com.

- *One-on-one training.* If there are sufficient resources available, this form of training is best, because the instructor can directly observe a trainee's comprehension of a topic and revise the training method on the spot. Unfortunately, the trainer is usually a senior staff person who has little time available for training.

- *Seminars.* A full-day seminar may cover substantially more than an employee's specific need, so review the exact contents of the program before sending an employee off for what may be a day of wasted training.

- *Specially-designed in-house classes.* If there is an in-house training department, request that it design a set of courses for the accounting department. This is not normally a cost-effective option unless the accounting department is large. It may be justified, however, if the topic covers an area in which the company occupies a particular niche in an industry; for example, a company that sells to the federal government may be justified in creating a comprehensive course that covers all accounting aspects of dealing with the government.

- *Trade association training.* There are a number of trade associations that publish a multitude of specialized books, many of which will be applicable to a company's training needs. Examples are the American Production and Inventory Control Society, the American Institute of Certified Public Accountants, and the Institute of Internal Auditors.

There can be considerable resistance to any type of training by employees, who often feel that they do not have enough time available for it. This issue is best dealt with by creating a formal training tracking system that records the hours spent on training, as well as any resulting test scores. These training results can then be incorporated into each employee's ongoing performance reviews. Also, if the accounting manager is responsible for updating each employee's monthly calendar of activities, she can add the training sessions to the calendars. Continuing attention to these issues is necessary to ensure that employees become fully trained in all areas of responsibility.

REVIEW WORK CAPACITY

Once all of the preceding issues have been implemented, it is still necessary to ensure that a sufficient number of personnel are available to complete the scheduled number of tasks. There are a wide variety of accounting tasks to be completed throughout the month, and they do not occur in a steady stream—requiring more people on some days than others. There are several ways to predict personnel capacity problems, as well as ways to deal with them. However, there are no analytically precise ways to ensure that the exact number of trained personnel are available as needed—capacity management involves a mix of experience, metrics, and guesses to arrive at an approximately correct solution.

One way to predict personnel needs is through metrics. For example, if one person can process 125 invoices in a day, then this number can be extrapolated over a larger number of invoicing clerks to determine how many of these people should be employed. Clerks who work on multiple tasks, as is the case in smaller organizations, are usually less efficient because they switch from task to task, requiring time to get up to maximum speed in each task. There is some setup time associated with each task—the longer the "production run" (in this instance, the number of invoices created), the smaller the amount of the setup cost on a per-invoice basis. Thus, a clerk who creates only a few invoices at a time must spread the setup cost of doing so across just a few invoices, which makes the per-unit cost more expensive. Consequently, proper metrics calculation requires a considerable knowledge of how the transactions are processed within an organization.

Another method for predicting personnel usage is the size of work backlogs. If work backlogs change, this is a sure sign of capacity overload or underload.

It is also possible to have employees form their own review group to ascertain the need for more or fewer employees. Though there is always some reluctance to recommend a cut in their own ranks, such a group can achieve surprisingly accurate estimates of projected capacity.

The final method of capacity problem analysis is the most common—to periodically review a trend line of overtime hours worked. Even if employees are paid on a salaried basis, they can still be asked to complete timesheets that reveal the amount of overtime worked. If the level of overtime is excessive, then it is probable that more employees are needed. However, by itself, this is an inaccurate method; hourly employees may take advantage of overtime pay situations, even if the amount of work on hand does not justify it, while the recorded time of salaried personnel tends to be quite inaccurate.

All of the preceding methods can be used, preferably in combination, to arrive at some reasonable estimate of future headcount needs. This is a difficult management task and should involve as many different methods of estimation as possible to provide the best input to the decision.

There are several ways to deal with capacity-related problems, which are as follows:

- *Accept the current workload.* The nature of accounting is that some transactions absolutely must be completed at certain times of the month, so there will be surges in transaction volume that require extra work. Employees can be told that overtime will be required during these intervals.

- *Share work.* If the accounting department is sufficiently compartmentalized, there will be situations where some employees are working less than others. If so, use cross-training so that more people can be brought to bear on a particularly large task.

- *Create a queue.* Determine which tasks can be safely delayed a few days or weeks. Then put these tasks on hold during periods of heavy transaction volume and complete them at a later date. The main problem is that the work in queue may never be done if the volume of work rises more than expected, which eventually results in a crisis situation until the work can be brought back under control.

- *Use temporary employees.* If there are predictable short-term surges in demand, then consider hiring temporary staff to assist.

However, their level of expertise will be problematic, so they can only be used for a limited number of tasks that require minimal training.

- *Hire new employees.* When all of the preceding options are no longer sufficient, it is time to hire extra employees.

Unfortunately, many managers do not first go through the earlier capacity management steps before hiring additional staff, so they end up with accounting departments where the workload is very uneven, and there are more personnel than are actually needed.

REVIEW CONTROLS

The accounting manager is largely responsible for establishing an adequate system of controls for the accounting transactions. Unless the controller is assisting with the formation of an entirely new company, there should already be a system of controls in place. In this case, the first task is to determine if there are any significant gaps in the existing control system. There are several ways to do so:

- *External auditor management letters.* Read the last few auditor management letters. These letters are issued by the company's external auditors, and point out various problems that the auditors uncovered during their audit.
- *Internal auditor reports.* If there is an internal audit staff, read any reports they have issued that describe control-related problems.
- *Schedule internal audit reviews.* If there is an internal audit staff, request a series of ongoing reviews of the various processes. It may take a long time for the audit staff to squeeze these requests into their schedule, but they can eventually provide relatively detailed appraisals of the key control systems.
- *Investigate transaction errors.* The accounting department is constantly fixing transactional errors of all kinds—incorrect customer

invoices, missing payments to suppliers, incorrect inventory counts, and so on. Each of these errors is a prime indicator of a possible control flaw, and should be used as such.

- *Hire a consultant.* The accounting department rarely has the manpower or the expertise to conduct an in-depth review of control systems. However, due to the requirements of the Sarbanes-Oxley Act, a large number of controls consultants have emerged who have the requisite level of expertise. Consider retaining one on a permanent basis to investigate controls throughout the company.

The preceding steps will give a controller a reasonable view of any control inadequacies, but does not provide a comprehensive view of the entire system of controls. It is extremely useful to document a complete set of controls for all key transactions, which can then be referenced whenever a systems change (which is what the remainder of this book is about!) is contemplated. The accounting staff will probably not have time to create such a document, so consider hiring a controls consultant to do so. This person should have considerable documentation skill, and also be able to spot missing or overlapping controls. For more information on this topic, please refer to the author's *Accounting Control Best Practices* book.

ESTABLISH FILING AND RETENTION SYSTEMS

Accounting transactions inevitably involve vast quantities of documentation, which an accounting department must store in such a manner that it can locate the documents within a reasonably short period of time. Many managers give the lowest possible priority to this task, since there is *usually* not a great deal of risk associated with an inadequate filing and retention system.

However, there are times when such systems can be invaluable. For example, if a government entity requires documentation of certain transactions (as is common for federal contracting arrangements) or if a company is subject to a lawsuit that requires the production of older documents, there may be the prospect of penalties or at least significant research costs to produce documents. Also, for more recent documents, such as those associated with transactions in the past year, there is a strong chance that the accounting staff will need ready access to the information. There are a number of techniques for establishing an adequate filing and retention system that do not require a great deal of staff time. A synopsis of several advanced systems follow; for more information, please refer to Chapter 10, Data Collection and Storage Systems:

- *Time clock storage.* Rather than record hourly employee time on cardboard time cards, have employees punch their time directly into a computerized time clock, which in turn is accessed by a local computer.

- *Direct entry through web site.* Have suppliers enter their invoices directly into the company's accounting system through its web site. Though this requires more effort by suppliers, it allows the company to completely avoid the data entry or subsequent storage of supplier invoices.

- *Electronic document storage systems.* As documents arrive in the accounting department, scan them into an electronic document storage system, from which they can be more easily accessed. In most cases, the original documents can be eliminated at once, thereby reducing a great deal of storage.

- *Data warehouse.* A company can design a centralized data warehouse, which pulls in selected information from the accounting and other systems, and stores it for various reporting purposes for a number of years. This is typically a very expensive option.

At a minimum, when a reporting year is completed, be sure to store the related documents in banker's boxes and clearly mark the contents on each box, as well as the document destruction date. Given the frequency of document review, it is generally best to keep the immediately preceding year's archives in a location relatively close to the accounting operation. Documents from years prior to that date can usually be safely stored in a secure location in a warehouse, since they will rarely be accessed again.

Some documents must be segregated from the regular accounting documents, and retained for a substantially longer period of time. Board minutes, title records, contracts, and similar documents should be retained in the most secure location, and be available for ready access. These documents are not usually stored in banker's boxes, but rather in filing cabinets, for easier access. The most critical documents should be retained in a fireproof filing cabinet. By segregating these documents, it is much less likely that they will be accidentally destroyed during the annual document destruction process.

A final thought is that the cost of on-site document storage is much higher than one might suppose: the costs of office rental space, storage cabinets, and fire suppression systems can add up. For a larger corporation with a lengthy history, the cost can be quite dramatic. Thus, it behooves the controller to actively minimize the total amount of documents stored, and particularly the amount of on-site storage.

IMPROVE THE DEPARTMENT

Once the accounting department is set up, there is the potential for an endless series of improvement projects. Being a department that handles large quantities of paper, the accounting area can rapidly become choked with it, which greatly impedes the flow of work. Also, certain employees are more comfortable with clutter than others, and so will allow unusual quantities of materials to proliferate. Managers can attack this problem by focusing on the ongoing elimination of the following items, all of which interfere with the orderly flow of work:

- Unneeded chairs, desks, filing cabinets, and carts.
- Unneeded computers, printers, phones, copiers, and fax machines.

- Unneeded posted items, such as outdated labor law posters, white boards, corkboards, and old messages on those boards.

- Excess quantities of office supplies at employee desks, such as printer paper, staplers, tape, and so on.

It is particularly important to search in all possible areas for these items: in corners and behind desks, and especially in drawers, cabinets, and closets, where such items tend to accumulate. To keep accumulation from occurring in these areas, consider removing all doors and drawers in the department. This means that all items are out in plain view, where they can be more easily monitored and therefore eliminated.

Whenever these items are removed from the accounting area, don't immediately throw them out or send them back to a storage area. Instead, pile them in a holding area that is readily accessible to the staff, so they have a few days to retrieve anything they really need. Then remove the remaining items from the holding area. It may be necessary to tag each item in the holding area, to identify where it came from and how long it should stay there until it is removed. If it is unlikely that an item in the holding area will be used within the next year, then don't even consign it to storage—instead, put it in the trash or donate it. Otherwise the storage area will become excessively cluttered.

There is the particular problem of what to do about those employees who persist in piling up vast quantities of paperwork and supplies in their work areas. One possibility is the complete rebuilding of a work area. This involves *completely* emptying out someone's work area; even removing computer equipment and related cables. Then clean the entire area, and only put the most necessary items back. All other items are removed to a location well away from the employee, who then spends a week deciding how many items that were shifted elsewhere are actually needed. This process will likely liberate a startling number of supplies, and also allow a great deal of paperwork to be filed away.

This process is by no means an annual event—think more in terms of a weekly or monthly review of the accounting area. Clutter increases constantly, so only continual attention will keep it in check.

In addition to clutter reduction, consider reviewing work flows within the department. To do so, create a map of the department, noting all cubicles, office furniture, and equipment. Then note on it the travel paths taken by employees for *all* activities and note their frequency of travel. A likely result will be the repositioning or removal of furniture and equipment. Also, if the department relies on high-capacity, centralized office equipment, such as printers, faxes and copiers, it will likely make more sense to acquire a large number of lower-capacity units to position very close to individual employees or small groups of staff. Further, the map will clarify which employees need to be clustered together, along with certain document storage areas. By making these changes, travel times within the department can be substantially reduced.

A key detractor from optimum workflow is the filing cabinet. It is a central source of documentation, and because it is extremely heavy, it cannot be moved. Instead, employees must travel to it—possibly many times over the course of a day. An alternative is to buy a number of office carts with wheels; employees load these carts with the files they will need for that day, and position the carts nearby in their work areas. The amount of travel time reduction can be astonishingly high.

The cubicle can be a considerable detriment to an efficient department, because it cannot be easily moved. Instead, swap out cubicles for desks, which can be easily reconfigured to match work flows. For example, group together the desks of the billing, cash application, and collection employees, so they can more easily discuss payment issues. If there is an increased need for more staff in this area, then simply move another desk into the group. If there is no way to avoid cubicles, then at least reduce their wall height, so that employees can more easily interact. High cubicle walls are the bane of employee communications.

Another issue that is frequently overlooked is the storage and replenishment of supplies. There is usually a central storage area for the department, but this is not always the best way to position supplies. For example, consider positioning printer and copier supplies right next to those items, so that users do not have to travel anywhere to find them. Also, assign replenishment responsibility to a single individual, and make sure that person uses a standardized checklist to update supplies on a daily basis. Otherwise, the department will lose valuable time searching for supplies that do not exist.

The result of these activities should be a reduction in the accounting department's required amount of square footage, which in turn results in less travel time within the department. Thus, a good metric for departmental improvement is either total square footage or (better yet) square footage per person. This metric can be taken to an extreme, since the staff could end up packed together like sardines in a can, so don't use it as an exclusive metric.

In summary, the accounting department can be continually streamlined to improve its efficiency. The top improvement considerations are to move high-usage items as close to the users as possible, and to remove anything that physically gets in the way.

SUMMARY

The steps described in this chapter for setting up the accounting department may seem quite obvious; yet the accounting manager who has gone through these steps and regularly revisits them is a rare one. The main reason why such basic organizational steps are not done is that they require a considerable amount of management time. Those managers who think they do not have time for these activities should revisit their own job descriptions; they will find that organizing the department is the primary task of a true manager—all other activities are secondary to ensuring that the department runs with the greatest possible degree of efficiency and effectiveness.

Chapter 2

The Sales Cycle

This chapter discusses the entire sales cycle. A typical sales cycle has an extraordinary number of controls related to separating various duties in the process, all designed to ensure that no one can manipulate reported sales volumes or steal incoming payments for receivables. As transactions wend their way from one person to the next, it is very easy for a transaction to be halted for any number of reasons, each of which contributes to slowing the overall process.

This chapter describes a typical set of sales and accounting transactions and shows how to improve the overall transaction speed while still maintaining a proper level of control. We then review the modified system for control weaknesses, and follow with a detailed cost/benefit analysis for several aspects of the modified system, which can be used as the basis for justifying conversion to the new system. The chapter concludes with a discussion of the most appropriate metrics and reports that can be used to support the system.

CURRENT SYSTEM

The traditional sales transaction begins when a customer purchase order arrives in the company mail room. The mail room staff forwards the purchase orders to the order processing clerk, who is normally located in the sales department.

The order processing clerk uses the information on the purchase order to create a sales order. The sales order is a revised version of the purchase order that is restated to fit the company's order processing system. For example, the sales order may include the company's own part numbers for the items being ordered. The sales order also includes check-off spaces for such tasks as credit approval, entry into the production schedule, and shipping.

The sales order then moves to the credit analyst, who is usually located in either the treasury or the accounting department. If a customer is already set up in the system with an unused line of credit, the sales order is approved. If not, the credit analyst reviews the customer's financial condition by examining its credit report and possibly (for really large orders) its financial statements. If the credit information is acceptable, the analyst assigns a credit amount to the customer, approves the sales order, and passes it along to the production control department.

The sales order enters into the logistics area, which falls outside the analysis of this chapter. After the parts have been picked from stock, ordered and received from a supplier, or produced, they are sent to the shipping department.

The shipping department reviews the sales order to ensure that the credit analyst has approved the order. If so, the parts are shipped. If not, the parts stay in the shipping department until the credit analyst releases the order. Realistically, credit approval should not have to be checked in the shipping department, because the sales order should not have been released into the production process unless credit was approved. However, it is useful to perform this last-minute check, to guard against an improperly released sales order.

When the order is ready to ship, the shipping department creates a bill of lading to go with the shipment. The shipment is listed in a shipping register that shows the sales order number, freight carrier name, ship date, and customer name. The process thus far is noted, with additional controls, in Exhibit 2.1.

The shipping department then forwards a copy of the bill of lading to a billing clerk in the accounting department, who compares it to the

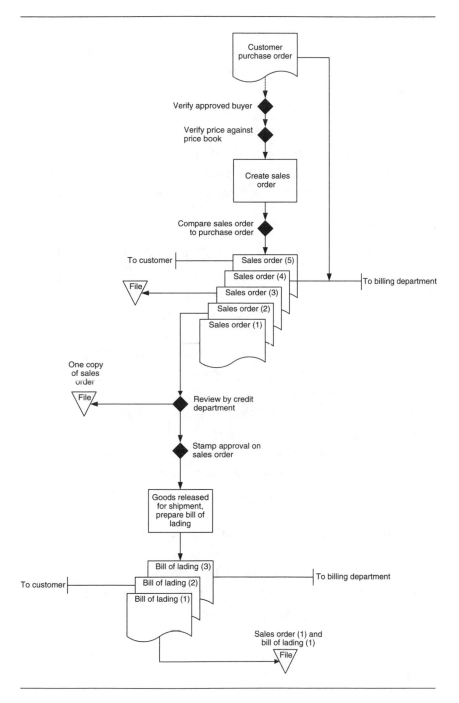

Exhibit 2.1 Typical Sales Transaction

sales order, customer purchase order, and any change orders. The accounting department applies any discounts based on purchasing volumes or special deals, and issues the invoice. If shipments are related to a long-term contract or are being sent to a government entity, the billing clerk may also need to review the related contract for pricing terms.

After the billing clerk completes all invoices for the day, she enters a batch total into a sales journal, which is later posted into the general ledger. The billing portion of the sales cycle is noted, with additional controls, in Exhibit 2.2.

The sale now transitions into a receivable. If payment is not received by the due date, the collections clerk calls the customer's accounts payable department to ascertain payment status. The collections clerk may have to send additional information to the customer, such as proof of shipment or a copy of the invoice. A variety of additional dunning activities may follow.

After the customer pays for the invoice, the accounting staff nets the received cash against the outstanding receivable. Normally, the customer payment includes a remittance advice that lists all the invoices being paid, which helps the accounting staff assign the payment to the correct invoices.

The value-added analysis shown in Exhibit 2.3 lists each step in the sales cycle and the time required to complete each step. A value-added item is considered to be one that brings the sales cycle closer to conclusion. This analysis shows optimistic wait times, while paperwork sits in employee mailboxes. A statistical analysis of an actual situation will reveal a small number of much longer wait times that will appreciably affect the completion time of the transaction.

Exhibit 2.4 presents a summary of the value-added analysis. It shows that only one-quarter of the steps bring the sales cycle closer to a conclusion; the remaining activities are related to moving paperwork between departments, which introduces the risk of lost or misinterpreted information.

The following section presents a revised system that includes numerous best practices. The intent is to streamline the overall

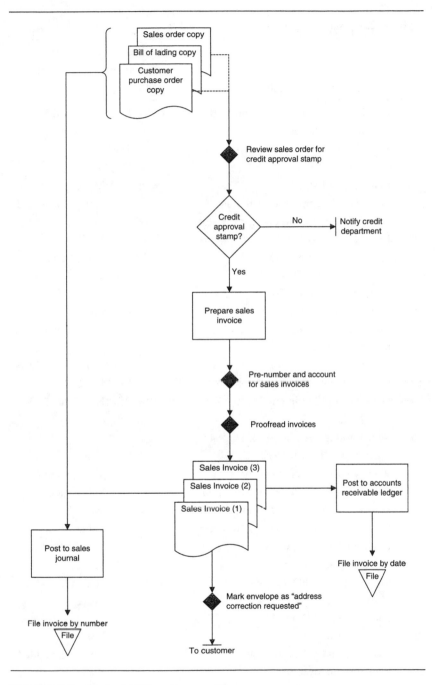

Exhibit 2.2   Typical Billing Process Flow

Exhibit 2.3   Sales Cycle Value-Added Analysis

Step	Activity	Time Required (Minutes)	Type of Activity
1	Receive customer purchase order (PO) in mail room	1	Value-added
2	Wait—accumulate until mail sorting is finished	15	Wait
3	Carry to mail slot	1	Move
4	Place POs in employee mail slots	1	Non-value-added
5	Wait for employee to check mail slot	120	Wait
6	Carry PO to desk	1	Move
7	Prepare sales order (SO)	10	Value-added
8	Wait while other POs in batch are converted to SOs	60	Wait
9	Move to photocopier	1	Move
10	Makes copies of PO	1	Non-value-added
11	Move back to desk	1	Move
12	File copy of PO	2	Non-value-added
13	Carry PO and SO to mail slot	1	Move
14	Place PO and SO in employee mail slot	1	Non-value-added
15	Wait for employee to check mail slot	120	Wait
16	Carry PO and SO to desk	1	Move
17	Review customer credit records	10	Value-added
18	Call customer to request information	5	Value-added
19	Wait for customer information to arrive	480	Wait
20	Review customer-supplied information	10	Value-added
21	If acceptable, approve SO	1	Value-added
22	Move to photocopier	1	Move
23	Make copies of PO and SO	1	Non-value-added
24	Move back to desk	1	Move
25	File copies of PO and SO	2	Non-value-added
26	Move to mail slot	1	Move
27	Place PO and SO in mail slot	1	Non-value-added
28	Wait for employee to check mail slot	120	Wait

(Continued)

Exhibit 2.3 Continued

Step	Activity	Time Required (Minutes)	Type of Activity
29	Move PO and SO to desk	1	Move
30	Write down order on production schedule	5	Value-added
[Production occurs and product is shipped]			
31	Move to copier	1	Move
32	Make copies of PO, SO, and bill of lading	1	Non-value-added
33	Move to desk	1	Move
34	File copies of PO, SO, and bill of lading	1	Non-value-added
35	Move to mail slots	1	Move
36	Place PO, SO, and bill of lading in mail slot	1	Non-value-added
37	Wait for employee to check mail slot	120	Wait
38	Move PO, SO, and bill of lading to desk	1	Move
39	Prepare invoice	5	Value-added
40	Wait for other invoices in batch to be prepared	60	Wait
41	Move to photocopier	1	Move
42	Make copies of PO, SO, bill of lading, and invoice	1	Non-value-added
43	Move to desk	1	Move
44	File copies of PO, SO, bill of lading, and invoice	1	Non-value-added
45	Move to mail room	1	Move
46	Mail invoice	1	Value-added
[Payment from customeris received]			
47	Mail room staff withdraws payments from letters	1	Value-added
48	Wait for all other mail to be opened	15	Wait
49	Mail room staff records check information on register	5	Value-added

(*Continued*)

Exhibit 2.3   Continued

Step	Activity	Time Required (Minutes)	Type of Activity
50	Move checks to mail slot	1	Move
51	Place checks in mail slot	1	Non-value-added
52	Wait for employee to check mail slot	1	Wait
53	Move to desk	1	Move
54	Contact customer about incorrect or unidentified payment amounts	10	Value-added
55	Post payment to customer account	5	Value-added
56	File invoice	1	Non-value-added

Exhibit 2.4   Summary of Sales Cycle Value-Added Analysis

Type of Activity	No. of Activities	Percentage Distribution	No. of Hours	Percentage Distribution
Value-added	13	23%	1.10	6%
Wait	10	18	18.52	91
Move	19	34	0.32	2
Non-value-added	14	25	0.27	1
Total	56	100%	20.21	100%

process sufficiently to reduce the overall time required to complete the sales cycle.

REVISED SYSTEM

This section addresses ways to speed up the sales cycle by such means as bypassing or altering selected steps, simplification, and automation. In a few cases, it is necessary to add steps in order to reduce the total time required by the sales cycle.

One solution to speeding up the sales cycle seems simple enough—just concentrate all the functions with one person (known as process centering) to avoid the move and wait intervals. However, doing so violates a number of key control issues. For example, a salesperson assigned responsibility for all of these tasks might grant excessive credit to *all* customers in order to maximize commissions. Or, someone who prepares and mails the invoice should not also be responsible for logging the payment into the accounts receivable system, since he could issue a dummy invoice to the customer for an excessive amount, and pocket the difference upon receipt. Finally, there is the risk that an employee could send a shipment to a fake company owned by the employee. For these reasons, other means must be found to improve the transaction speed.

One cycle-time compression option is to *pre-approve customer credit*. Nearly all companies allow their salespeople to pursue customers and procure orders before checking on the credit of the customers. This leads to excessive pressure on the credit analyst, who must now collect credit information about the customers and make decisions about the amount of credit (if any) to be extended. Meanwhile, pressure is brought to bear on the credit analyst by the salesperson, since the latter wants to hurry through the purchase orders that translate into commission income for the salesperson. Overall, the situation causes orders to be delayed, since it takes time to procure credit information and reach a decision, and can lead to bad credit decisions, since the sales department is determined to book the orders. An alternative approach is to team with the sales department in advance, go over the credit records of potential customers, and assign them credit before the salesperson ever makes a sales call. This keeps credit granting from being a bottleneck in the sales cycle, and also keeps the company away from potential customers with bad credit. The best way to sell salespeople on this change is to implement a pilot project with a senior salesperson and track the actual time reduction for the transactions of that salesperson's customers. Once the pilot results are made public, the other salespeople should be more willing to accept the procedural change.

Some companies resist the use of pre-approved customer credit. If so, a reasonable alternative for high-speed processing of credit applications is the *credit decision table*. This is a simple Yes/No decision matrix based on a few key credit issues. An example of how a decision table might work is as follows:

1. Is the initial order less than $1,000? If so, grant credit without review.

2. Is the initial order more than $1,000 but less than $10,000? Require a completed credit application. Grant a credit limit of 10% of the customer's net worth.

3. Is the initial order more than $10,000? Require a completed credit application and financial statements. If a profitable customer, grant a credit limit of 10% of the customer's net worth. Reduce the credit limit by 10% for every percent of customer loss reported.

4. Does an existing customer's order exceed its credit limit by less than 20% and there is no history of payment problems? If so, grant the increase.

5. Does an existing customer order exceed its credit limit by more than 20% or there is a history of payment problems? If so, forward to the credit manager for review. Use the same credit granting process listed in step 3.

6. Does an existing customer have any invoices at least 60 days past due? If so, freeze all orders.

This approach sets up clear decision points governing what actions to take for most situations, leaving only the more difficult customer accounts for further review.

Yet another way to streamline the credit granting process is to *automatically grant all new customers a minor credit line*, and then review the situation in more detail after some time has passed and more internal information has been compiled regarding customer ordering and payment habits. This system allows the credit staff to focus its attention on larger credit requests. This practice is most useful where there are many new customers, when the credit department has a significant work backlog, or when company products have such a high gross margin that a few bad debt losses will not cause a significant drain on company resources.

There may be several invoice iterations if the company uses a highly complex pricing model. For example, a product may include an array of separately–priced options, resulting in a multitude of possible price totals. There is a strong possibility that some of these billings will be incorrectly compiled and invoiced, leading to customer dissatisfaction and delayed payments. A difficult improvement is to *simplify pricing*. This calls for the cooperation of the sales department, which may object to fewer pricing options. The solutions used by many car companies is to offer many options, but to cluster those options into just a small number of available configurations, so that there are actually only a few price points associated with each vehicle.

When a company issues a large, complex invoice containing multiple line items, there is an increased likelihood that its customers will hold up payment of the entire invoice while arguing over a single line item. To avoid this, *print a separate invoice for each line item.* By doing so, a customer will only hold up payment on a single line item, while paying for all other items over which it has no issues. This can cause a major increase in invoicing volume, so only use it for what would otherwise be truly complex, large-dollar invoices. The only customer complaint arising from this approach is that they can be buried under large piles of invoices. This can be ameliorated by clustering all of the invoices in a single envelope to each customer, rather than separately mailing each one.

Another method for reducing processing time is to *create an invoice at the point of delivery.* This eliminates the time needed to create an invoice at the company location, mail it to the customer, and go through any invoice re-issuances that are caused by disputes over the amount received. Instead, the delivery person can have all relevant information stored in a portable computer, and print an adjusted invoice after the customer's receiving department inspects the shipment. This is only an option when the company has its own in-house freight carrier, since third-party carriers are not authorized to modify and deliver invoices.

A customer may set up new suppliers in its computer system using default payment terms that it applies to all customers. To ensure that customers use the correct payment terms, *issue a payment procedure* with the first invoice. This procedure should list the company's payment terms, note contact information, the process for returning goods, and how to claim a credit. It can also include a form for payments via the Automated Clearing House (ACH) Network.

When a customer receives an invoice and has a question about it, who does he call? The invoice usually only contains a mailing address, so the customer elects to wait until a collection person calls, which delays payment. The solution is to *clearly state contact information on the invoice*. This should be delineated by a box and possibly noted in bold or colored print. If the billing staff is large, it may not be practical to put a specific contact name on the invoice, but at least list a central contact phone number.

Sometimes, the layout and content of an invoice can be one of the largest payment impediments, because customers cannot figure out what they must pay, when it is due, or where they should send payment. Here are some tips for *improving the invoice format*:

- *Add credit card contact information.* This gives customers another payment option, and so may increase the odds of prompt payment.

- *Clearly state the payment due date.* Clearly state the actual invoice payment date, rather than describing such payment terms as "net 30," which otherwise gives the customer some leeway in calculating the payment date.

- *Eliminate information not needed by the customer.* The customer does not care about the initials of the salesperson on their account, or the job number, or the document number (which is sometimes listed in addition to the invoice number). Strip out all information that is not specifically needed by the customer to pay the invoice.

There is usually a significant emphasis on producing invoices as soon after the completion of service or product delivery as possible, on the grounds that this compresses the sales cycle. While correct, it can also result in the issuance of incorrect invoices, which ultimately results in a much longer sales cycle, due to the delays required to subsequently correct and re-issue invoices. The solution is to *selectively proofread the largest-dollar or most complex invoices.* The assigned proofreader should not be the person who initially created the invoice, and so has an independent view of the situation and can provide a more objective analysis of invoice accuracy. This does add an extra step to the invoicing process, but ultimately shortens the entire sales cycle.

Larger customers with more automated accounting systems may not want to receive a paper-based invoice at all, since they must manually process it. If so, they may demand that the company *manually enter the invoice information directly into the customer's computer system*, probably through an Internet site. If so, do it! This certainly requires more work on the company's part, but greatly reduces the risk of non-payment. Better yet are those customer systems allowing for the entry of a number of supporting documents, such as the authorizing purchase order and shipping documentation. The more information that is available to the customer, the greater the likelihood of timely payment. If the company issues a large number of invoices to the customer, it may be possible to arrange for an automated feed, but don't count on it. This is a clear case of adding work into the sales cycle in order to reduce the overall length of the cycle.

A company that ships to a customer on a just-in-time basis may be able to *eliminate all invoices* by having the customer pay based on the number of the company's products used in the customer's final product. In essence, the company frequently ships a small number of its products to the customer; the company is the only supplier of that product to the customer. When the product arrives at the customer's production facility, it is immediately included in the production process. When the customer's product is completed, the customer's accounts payable system can calculate the amount due to the company based on the number of its parts in the customer's finished product. This system has no need for invoices. Instead, payments will be sent to the company periodically; the remittance advice attached to the payment will list the customer's purchase order number as a reference instead of the usual invoice number. In short, the invoice creation process can be avoided entirely if the company's customers pay based on their production records. This solution is usually only available from very large customers who have invested heavily in advanced accounting systems.

The elimination of invoices presents some tricky implementation issues. For example, some accounting software requires that invoices be printed before they will process sales transactions, so a company must print and then discard them. Another issue is that it can be difficult to match the company's shipment records to customer payments, since the customers no longer match payment amounts to invoice numbers on the remittance advice; this is normally handled by tracing back through the purchase order number instead.

It is also useful to consider the timing of an invoice. It is possible in some situations to *issue invoices before the invoice date*. This is possible when a company is billing recurring invoices in highly predictable amounts, as is the case with subscriptions. The accounting department sets the invoice date forward to the legally agreed-upon billing date, creates the invoice, and mails it early. By doing so, the customer has more time in which to process payment, thereby improving the odds that it will be paid on the due date. The main difficulty is ensuring that the date of the invoice is actually shifted forward into the correct accounting period; otherwise, the company is recognizing revenue too early.

Customers move to new locations all the time and sometimes do not notify their suppliers of this change. The result is invoices going to old addresses, and therefore delayed payments. There are several methods available for reducing the number of invoices returned by the Postal Service. First, *mark all envelopes as "Address Correction*

Requested." If a customer has filed a forwarding address with the U.S. Postal Service, the Postal Service will not only forward the mail to the new address, but also notify the company of the new address. Second, *have the company's mail room staff expedite all returned mail back to the accounting department*. The accounting staff should give this mail the highest priority in researching the correct address, entering changes in the accounting database, and reissuing the invoice. Finally, have the *sales staff review contact information for upcoming recurring invoices*. It is likely that the sales staff stays in constant contact with customers for whom there are long-standing billing situations, and so will have a good idea of any contact or address changes.

A way to reduce the workload of the billing staff is to *eliminate month-end statements*. Many accounts payable departments do not review these statements, and simply throw them out. Furthermore, a periodic statement is not an invoice, so a customer cannot pay the company from the statement anyway—it must call the company and request an invoice for any items listed on the statement that are not in its records. Very few payables departments have the time to follow up on statements in this manner. Thus, it is best to eliminate the printing and mailing of these statements.

The customer's accounting department may lose an invoice. This is because the payables staff may send invoices out to supervisors for approval, and the supervisor loses or delays it. If the invoice is missing, the customer's payables department has no way of knowing about it until a collections call is made for an overdue invoice; by the time a replacement invoice has been sent, even more time will have passed. To avoid this problem, the collections staff can *send invoice reminders prior to the invoice due date*. The most efficient way to do so is to e-mail the reminder, with an attachment containing an invoice copy. This chore can be automated, so that the accounting software issues the reminder a specific number of days prior to the due date. For larger invoices, the collections staff may want to do this manually, to be absolutely certain that the customer has received and processed the invoice. The company's collections staff may protest this extra work, since they will claim that there is not enough time to make collection calls about overdue payments, much less to customers whose payments are not yet due. A possible solution is to temporarily use additional collections staff to contact the most troublesome overdue accounts until the full impact of the early contacts takes effect.

Once invoices have been issued, the next emphasis is on collecting cash. The simplest way to do so is an automated bank account deduction, also known as an *ACH debit*. In this case, a customer authorizes the company to debit its bank account in the amount of any (or specific) invoices, on a pre-determined payment date. Customers usually only agree to this approach if invoices are for the same amount each month, and are not subject to much debate (as would be the case with a monthly rent payment). In other cases, customers prefer to retain the right to argue over the amount of an invoice, which is difficult to do when the company is removing the funds from the customer's account in any event.

Customers sometimes have difficulty delivering payments to a company; they may send it to the wrong address, or to whomever they know at the company. Also, even if a check safely arrives at the company, it may be delayed before finally being deposited. To accelerate the sales cycle in this area, *have customers send their payments to a lockbox*. This is a mailbox to which the company's bank has access, and from which it extracts mail each day and deposits all checks contained therein. This can potentially cut several days from cash processing. The company can usually access a website each day to see images of all checks that were deposited on the previous day, and which it uses to record cash receipts within its accounting system. A variation on this approach is *lockbox truncation*, whereby a company has a check scanner on its premises, and which it uses to scan received checks and transmit an electronic deposit transaction to its bank.

The effort of collecting open receivables can be reduced by *immediately recording cash receipts*. Otherwise, the collections staff will waste time trying to collect money that has already been received. This requires the allocation of sufficient staff time to ensure that all cash is promptly recorded against open receivables.

In addition to fast cash application, a company must also be mindful of *unapplied cash*. This is typically check payments for which remittance advices are missing, or for which payments do not match any known invoices. The sales cycle will not end for the invoices offsetting these checks until such time as staff time can be found to research them and determine the most appropriate allocation. Thus, a best practice is the *immediate review of unapplied cash*. Though this requires the continuing attention of senior clerical staff, it is still critical to "close the loop" on payments and clear the related receivables from the company's records.

A flawed or indifferent collections system can delay the sales cycle to an extraordinary degree, with some invoices remaining unpaid far past their due dates. In most companies, there are not enough collections staff available to track down the reasons for non-payment, so some invoices that could otherwise have been collected will be written off, or passed to a collections agency. When there are not enough available staff, an excellent management technique is *collection call stratification*. The concept behind this approach is to split up, or stratify, all of the overdue receivables and concentrate the bulk of the collection staff's time on the largest invoices. For example, a collections person can be required to contact customers about all high-dollar invoices once every three days, but limit contacts regarding small-dollar invoices to once every two weeks. By using stratification, a company can realize improved cash flow by collecting the largest dollar amounts sooner. The downside of this approach is that smaller invoices will receive less attention and therefore take longer to collect, but this shortcoming is reasonable if the overall cash flow is improved.

Another way to improve the collection of open receivables is to *acquire or create a collections software package*. This is a computer program that is tied to the due dates on the company's accounting software, and which tells the collections department when an invoice is overdue. The collections staff can type contact information into the collections software, which is then available the next time someone contacts the customer. The system allows the collection staff to make more calls, since they are not wasting any time searching for information about overdue accounts—the correct information is presented to them by the system.

This level of automation is incorporated into some higher-end accounting software packages, or as a commercially-available package, such as GetPAID. This software requires a customized interface to the existing accounting software, which it uses to extract information about customers and unpaid invoices. The software notes all contact information on a single screen, including the results of previous conversations with each customer. If a fax router is available, a collections person can even send an invoice copy to a customer without ever leaving his seat. The system will also automatically fax or email reminder messages to customers when payment due dates arrive, and can prioritize collection calls by time zone, so that those customers in time zones that are outside of normal business hours will not be called. This type of solution can be a major efficiency improvement for a larger company facing significant collection problems.

A more limited subset of a collections software package is a comprehensive *database of customer contacts* that can be used to collect payments that would otherwise be stalled. This list should not just include members of a customer's accounts payable staff, but also anyone higher up in a customer's organization. There should be a personal connection with any person on the list, so the individual will be more likely to assist the collections staff with an occasional request. The database will likely require considerable input from the company's sales staff, which is most likely to have numerous customer contacts. The collections staff should use the list with care, since going outside of a customer's accounts payable chain of command can have adverse long-term ramifications.

Bad debts will occur despite a company's best efforts. However, the accounting, credit, and sales staffs can learn from these bad debts by conducting a *bad debt post mortem*. This meeting includes the credit and sales managers, as well as those salespeople whose accounts caused the bad debts. This group should discuss the reasons for the bad debts, what systemic changes can be implemented to reduce the likelihood of their recurrence, and assign responsibilities for any actionable items. This creates an excellent feedback mechanism for reducing the problems leading to bad debts, while giving all involved personnel an education about why bad debts occur.

Similarly, there should be a *payment deduction database*. This is used to accumulate the reasons customers give for taking deductions from their payments. In many cases, these reasons reflect problems originating within the company, such as faulty products, incorrect order processing, product damage caused by incorrect shipment packaging, and advertising deductions taken for deals not clearly defined by the marketing staff. Managers can then access this database to ascertain which problems are causing the most collection difficulty, and work on solving the largest-volume deduction problems first. It is also useful to link this database to a workflow management system, so that problems can be routed to the correct person, with automated progress reports being sent to managers. The system should also be able to shift the action routing to an alternative party in case action is not taken by a predetermined date, thereby ensuring that action is taken to reduce deduction-related problems.

The accounting staff should also issue *accountability reports for bad debts*. It is entirely possible that an inordinate number of bad debts can be traced to a specific salesperson, production batch, or some other cause. If so, create a report that targets the specific problem, so that management can concentrate its efforts on error remediation. Similarly, issue *collection effectiveness reports* that describe which collections staff are the most effective in terms of days' sales outstanding. This later report can be subject to considerable misinterpretation, if a collections person is assigned an especially difficult customer, so use it with care.

In summary, the sales cycle can be improved by bypassing or altering some steps, work simplification, and the use of automated systems that are linked to a central database of accounting information. A very simplified flowchart of the revised sales cycle is shown in Exhibit 2.5. In particular, note the absence of paper documents in most parts of the revised process.

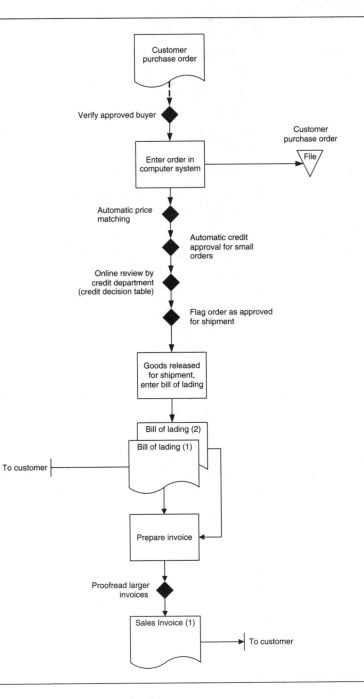

Exhibit 2.5   Improved Sales Cycle

REVISED SYSTEM—ELECTRONIC ORDERS

The sales cycle becomes even more compressed when customers place orders through electronic forms over the Internet, or through Electronic Data Interchange (EDI) systems. In both cases, customer orders are automatically inserted into a company's computer system. By doing so, there is no need for overwhelming controls over the accuracy of records, since the input system should have automatically screened for errors. For the same reason, there is no need to review price matching, nor any special terms. Further, if a customer pays with a credit card at the time of order placement, there is also no need for a credit review. Consequently, electronic orders can breeze through a company's order entry system without delay. The result is shown in Exhibit 2.6, which is primarily notable for the number of controls that are no longer used.

CONTROL ISSUES

There are many control issues related to the sales cycle. For example, if the controller were to eliminate controls related to the granting of customer credit, the risk of shipping products to nonpaying customers would increase. Similarly, removing controls over the frequency of collection calls would probably increase the number of bad debts. This section shows what would happen if various sales cycle controls were removed, and suggests possible solutions to the resulting control problems.

- *Documents are not manually transferred between departments.* Having the credit department manually stamp approval on an incoming customer order is a control over the granting of credit to customers. If the control were removed, orders could be shipped without the approval of the credit department, which might result in shipments to unqualified customers who cannot pay. To avoid this issue, the computer system can freeze a customer order if the credit department does not check off an approval flag on the order.

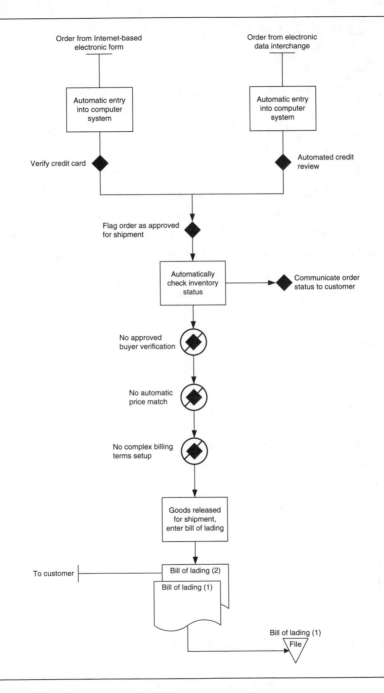

Exhibit 2.6 Sales Cycle When Electronic Orders are Used

- *Customer credit is preapproved.* Reviewing a customer's credit at the time the order arrives is a control over the *current* credit situation of the customer. Preapproving customer credit means that some customer credit information may change for the worse between the time the credit is approved and the customer places the order. However, the credit rating of the typical customer does not change very frequently; if bankruptcy occurs, it is usually the result of a long-term slide into insolvency that can be predicted based on a periodic review of the credit records. Consequently, a review of customer credit information just a few times a year is generally sufficient to cover any changes in the financial condition of a customer. This review can be augmented by subscribing to a credit reporting agency, which will automatically email the credit department with notifications of worsening customer credit scores.

- *Month-end customer statements are eliminated.* Sending a monthly statement to customers is an additional means of ensuring that they will pay their bills. Eliminating the statement means that customers are receiving one less notification of the level of their indebtedness to the company, which may reduce the likelihood of payment. However, the company can replace this passive contact with a more active contact that is more likely to result in payment, such as contacting the customer prior to the invoice due date to discuss any possible payment issues. This actually strengthens the control, since the collections staff will become aware of collection problems more quickly than if only monthly statements were used.

- *Invoices are eliminated.* Sending an invoice to a customer ensures that the customer has received notification of payment, and will therefore be more likely to pay. Eliminating the invoice presents a serious risk that payment to the company will not be forthcoming. It is still very unusual for customers to pay suppliers with no invoice, so the two parties should carefully review all aspects of delivery and payment in advance, and thoroughly test the proposed system. Also, there should be a fallback set of delivery information that can be used to justify a payment claim. Further, such payment systems are usually based on long-term contracts, so include a claims resolution procedure in the contract.

- *Invoices are delivered by the delivery person.* Segregating the invoice generation task from the shipping staff is designed to prevent the shipping department from keeping a portion of each shipment, while still billing the customer for the full balance. If the invoice generation task is given to the delivery driver, it is possible for invoices to be fraudulently altered. To mitigate this risk, predetermine the amount of the invoice, based on the shipping quantity loaded onto the truck. If the driver's invoice differs, then the accounting staff can investigate further. Also, these variances can be tracked by individual driver over time, to see if there are any consistently negative variances for specific drivers that may indicate pilferage. Further, the customer acts as a natural control, since it is presumably comparing the delivered quantity to the amount on the invoice.

All of the preceding changes result in control problems, but the issues are reduced or eliminated by the imposition of new practices that yield the same or stronger control during the transaction cycle. These changes are minimal when compared to the long list of sales cycle controls that are still necessary in nearly all corporate environments. The following controls should be maintained to avoid the potential increase in risk associated with their elimination:

- *Divide responsibility.* The majority of tasks in the sales cycle must still be divided among a large number of people to avoid the risk of modification of sales information by unethical employees. This means that duties should be separated for the preparation of sales orders, credit approval, issuance of merchandise from stock, shipment, billing, approval of sales returns and allowances, and authorization of bad debt write-offs.
- *Use periodic audits.* Many controls do not have to be performed continually. Instead, they are conducted on a spot basis; these are

ideal tasks for the internal audit group. The following items are representative of this type of control:

- *Check invoice prices against price lists.* Employees may charge customers the wrong price for a variety of reasons. The internal audit team can compare charged prices to official prices periodically to ascertain the extent of issues in this area. Problems most frequently arise when the pricing structure is complicated—when prices are based on volume or special deals.

- *Confirm receivable balances.* A confirmation request to a customer may reveal that an employee is not correctly applying cash received. However, a confirmation may not be returned by the customer, who is too busy to attend to the request.

- *Confirm that credit approval is required to ship.* By various means, customer orders may bypass the credit approval process and be shipped without approval. The internal audit staff can examine the paperwork of company orders to discover if this is occurring.

- *Differentiate between consignment and regular sales.* Products can be shipped to customers and still be the property of the company if there is a consignment sales agreement with the customer. This has an effect on the reported revenue number. The internal audit team can discover this problem by confirming sales agreements with customers and by reviewing internal sales agreements and customer purchase orders.

- *Look for bill and hold transactions.* Products can be billed to the customer but retained at the company's location. This causes serious revenue recognition concerns, since products are supposed to be shipped before revenue is earned. The audit staff can compare billings to the shipping log to find this problem, as well as review finished goods in the company's warehousing facilities.

- *Confirm that invoices are always issued.* If there were collusion between the shipping and accounting departments, products could be shipped to a false address without an invoice being generated, resulting in employee ownership of the product. This can occur if products are shipped without an entry in the shipping log, which does not require collusion. The internal audit team can look for this problem by examining inventory variances, comparing bills of lading to the shipping log, and comparing bills of lading or the shipping log to invoices issued.

○ *Compare bad debt write-offs to cash received.* An employee with control over bad debt write-offs and cash receipts can write off accounts that are collectible and then pocket the money when it arrives. The internal audit team can find this problem by comparing bad debt write-off amounts to the cash receipts record maintained by the mail room, or by reviewing the receipts record related to the company lockbox.

○ *Confirm that invoices are issued in the correct amount.* An employee with control over both the invoicing and cash receipts functions could invoice a customer more than the correct amount, record the correct amount in the sales journal, and pocket the excess amount of cash when it arrives. The audit team can find this problem by reviewing cancelled checks to ensure that they are being cashed into the company's account, comparing individual invoice copies to the sales journal, and independently calculating the correct amount of invoices based on the shipping log and comparing this to the amounts actually invoiced.

• *Require approval of discounts, returns, and allowances.* It is possible for an employee to issue an unauthorized discount to a customer in exchange for a kickback. Alternatively, this may be a way to collect the amount of the discount when payment is received from the customer (though this requires collusion with the person recording cash receipts or the combination of the two functions with one person). This problem is avoided if supervisory approval is required in order to process a discount, return, or allowance.

• *Review the period-end cutoff.* A company may report an incorrect amount of revenue in a period if items are billed that have not yet been shipped. A company can avoid this problem by comparing invoice records to the shipping log at the end of each reporting period and moving invoices into the next reporting period if they have been issued prematurely. This procedure also discovers items that have been shipped but not billed.

In addition, there are several system changes that can improve the flow of paperwork through the sales cycle, resulting in fewer errors. They are as follows:

- *Automate the shipping notification to accounting.* The most damaging error in the sales cycle is to not invoice the customer. This happens when there is a weak link between the shipping department and the accounting department; shipping information does not necessarily reach the accounting staff, which therefore issues no invoice. This can occur when the shipping log is not sent to the accounting staff at all (forgetfulness on the part of the shipping staff), when the shipping information is lost in transit between the shipping and accounting departments, and when the accounting department neglects to create the invoice. The weak link in the shipping notification can be strengthened by automation—the shipping staff types shipment information directly into a computer terminal, and the information automatically appears in the accounting database. If there is a projected ship date on a customer order, then the accounting staff can create a report such as the one shown in Exhibit 2.7, listing orders with overdue ship dates or no invoices. This report flags orders that may have been shipped but not invoiced.

- *Credit flag required.* Another error is the shipment of products to customers whose requests for credit have been turned down. This is caused by a breakdown in procedures, because no orders should be shipped until credit approval has been given. To resolve it, have the credit department check off a flag in the computer system to indicate credit approval, and then alter production and picking reports, so that unflagged orders do not even appear; what the warehouse staff does not see on its packing reports cannot be picked,

Exhibit 2.7 Orders in the System with No Invoices

Customer	Order No.	Amount ($)	Projected Ship Date
Able Phone Systems	0001123	14,321	10/10/2009
Bosch Tissues Inc.	0001127	9,267	10/08/2009
Cummins Lighting Co.	0001097	8,401	09/30/2009
Dorfman Thermometers	0001092	3,232	08/31/2009
Englehart Shipping Lines	0001199	9,461	10/03/2009
Finley Truck Lines	0001204	1,001	10/02/2009
Grimm Kid's Games, Inc.	0001207	2,003	09/22/2009

and the production management team will not schedule an order for production for the same reason.

- *Dedicated order receipt systems.* When a customer sends in a purchase order, it may be lost by the company before it even enters the sales cycle, resulting in a lost sale or at least a dissatisfied customer. An order can arrive in the mail, over the phone, at the fax machine, or by hand. The order can be lost at any of these points of arrival. To solve the problem, the fax machine can be set up in the order entry department, thereby reducing the chance of an order's being lost by someone outside of the department. The fax machine should have a considerable amount of memory, so that documents are stored in its memory even if there is no paper in the machine. If an order is delivered by phone, the reception staff must route the call to a central order entry phone number that has a voicemail backup, so that the order can be stored in voicemail even if no one is there to take the call. If the order is delivered by mail, the company can install a scanner in the mail room, so that purchase orders are entered into the system prior to delivery of the paper document to the order entry department. This means that a digital record is kept of the purchase order, even if the paper record is lost. Finally, if an order is delivered by hand, it should be routed through the mail room first, so that the document can be scanned (as just noted).

In summary, most controls related to the sales cycle cannot be changed without seriously weakening a company's control structure.

However, using technology or alternative processing steps can reduce the number of control steps used in a typical transaction.

COST/BENEFIT ANALYSIS

This section demonstrates cost/benefit analysis for pre-approval of customer credit, the elimination of month-end statements, collection software, calls prior to invoice due dates, and the elimination of all invoices. The expected revenues and expenses used in these examples will vary considerably from a company's actual situation, but the format used is a good framework for a realistic cost/benefit analysis. Examples are as follows.

PRE-APPROVE CUSTOMER CREDIT

Adi Pinkerson, the CFO of the CD Warehouse distribution center, is concerned about the time spent qualifying CD retail stores for credit after orders have been placed. She notes that many of the company salespeople are disappointed when they pursue potential customers and close deals, only to have their credit turned down by the credit department. She decides to look into switching over to pre-approved credit before the salespeople make any sales calls. The sales department estimates that 2 hours of the sales manager's time will be needed each month to collect lists of potential new customers from the sales staff and pass along that information to the credit department. The sales manager earns $85,000 per year ($40.87 per hour). Each monthly list will contain an average of 40 new customers. The credit staff must take an hour to collect and review credit information about each potential customer. The average hourly pay for each collections clerk is $16.00. Ms. Pinkerson estimates that this advance work will keep the credit department from turning down an average of five customers per month while saving the sales staff an average of 8 hours to secure each of the five orders. The average sales employee earns $60,000 per year ($28.85 per hour). Should customer credit be preapproved?

The savings come from having fewer credit checks to turn down and a significant decrease in the work of the sales department, while the additional cost primarily comes from having to pre-qualify a large number of potential customers who may not even place an order with the company. The analysis follows:

Cost of Pre-Approving Customer Credit

Sales manager time/month to collect information	2 hours
Months/year	× 12
Hours/year	24 hours
Sales manager cost/hour	× $40.87
Total sales manager cost/year	$981
Credit clerks time/month to review customer credit	40 hours
Months/year	× 12
Total hours/year	480 hours
Credit clerk cost/hour	× $16.00
Total credit clerk cost/hour	$7,680
Total cost/year of pre-approving customer credit	**$8,661**

Benefit of Pre-Approving Customer Credit

Time/month saved by avoiding credit reviews	5 hours
Months/year	× 12
Hours of credit reviews avoided	60 hours
Credit clerk cost/hour	× $16.00
Total savings/year from avoiding credit reviews	$960
Customers turned down/month for bad credit	5
Months/year	× 12
Customers turned down/year	60
Salesperson time to secure an order	× 8 hours
Salesperson time/year to secure bad credit orders	480 hours
Salesperson cost/hour	$28.85
Total cost/year of unnecessary sales calls	$13,848
Total savings/year pre-approving customer credit	**$14,808**

With costs of $8,661 and savings of $14,808, it is evident that the change should be made. The key reason for the change (in terms of cost/benefit) is that it saves the time of the sales department, whose employees are more expensive per hour than those of the credit department.

ELIMINATE MONTH-END STATEMENTS

Ms. Bentley, the controller of Ancient Autos Warehousing, is concerned that month-end statements are not generating any savings for the company, which sends out 1,250 statements per month to its customers. After a three-month test period when no month-end statements were sent, the company's days receivables measure increased from 50 days to 51 days, an increase of 2%. The company charges each customer an average of $200 per month to store an expensive car. The company earns 9% on its investments. Ms. Bentley spends $250 per month on statement forms and another $1.25 in labor and mailing costs to send out each statement. Should the company eliminate month-end statements?

The cost savings come from the materials, postage, and labor costs associated with a period-end statement mailing. These savings are offset by a slight worsening of the company's receivables balance. The analysis follows:

Cost of Deleting Period-End Statement Mailing	
Number of customers	1,250
Average balance/customer	× $200
Total average receivables balance	$250,000
Increase caused by lack of statements	× 2%
Amount of increase	5,000
Investment rate of interest	× 9%
Interest cost of excess receivables	**$450**
Benefit of Deleting Period-End Statement Mailing	
Cost/month of statement forms	$250
Months/year	× 12

(*Continued*)

Cost/year of statement forms	$3,000
Numbers of customers	1,250
Months/year	× 12
Invoices mailed/year	15,000
Mailing cost/statement	× $1.25
Statement mailing cost/year	$18,750
Total savings/year from deleting mailing	**$21,750**

The cost of eliminating monthly customer statements is $450, versus printing and mailings savings of $21,750. This cost/benefit analysis indicates that customer statements should be eliminated.

DELIVER INVOICE WITH SHIPMENT

John Walker, president of Walkers Deluxe, manufacturers of top-quality hiking shoes, wants to reduce the time needed to invoice customers. One possibility is to have the company's delivery drivers carry portable computers that are loaded with invoicing information. The CFO, David Slog, reviews the costs and benefits of installing such a system. Each of the company's ten delivery drivers would be equipped with a portable computer costing $2,200 and a $300 printer. In addition, each driver will need an hour of training. The average driver wage is $19 per hour. By delivering the invoices at the same time as the products, Mr. Slog estimates invoices would be paid two days sooner than usual. Since the current days' receivables measurement for the company is 41 days, this would be a drop of 4.9%. Walkers Deluxe has an average outstanding receivables balance of $3,450,000 and earns 8.5% on its investments. Should this system be adopted?

The cost savings are from reducing the company's receivables balance. These savings are primarily offset by the cost of computer hardware. The analysis follows:

Cost of Delivering Invoice with Shipment

Cost of portable computer	$2,200
Cost of printer	300
Computer hardware cost	$2,500
Number of drivers	× 10
Total computer hardware cost	$25,000
Number of drivers	10
Training time required/driver	× 1 hour
Total training time required	10 hours
Driver pay/hour	× $19
Total cost of training	$190
Total cost of delivering invoice with shipment	**25,190**
Benefit of Delivering Invoice with Shipment	
Current receivables balance	$3,450,000
Reduction in receivables with new system	× 4.9%
Amount of receivables reduction	$169,050
Corporate return on investments	× 8.5%
Savings from delivering invoice with shipment	**$14,369**

The cost of delivering invoices with the shipment is $25,190, which is counteracted by savings of $14,369. With a payback of 1.75 years, it appears reasonable to proceed with the project. However, be wary of the replacement interval of the portable computers, which may not last long in the field.

INSTALL COLLECTION SOFTWARE

Dudley Anderson, the collections supervisor at Smith & Smith Gunsmiths, goes to a collections conference and hears about collection software. He wants one. As the controller, you review the marketplace and find that such a system must be designed, not bought, since the only way to purchase a mid-range system is as part of a complete accounting package. It is too expensive to switch accounting packages just for this system. To build one in-house, the programming staff requires 2,200 hours for design, programming, testing, and documentation. The average programmer earns $30 per hour. In addition, a total of 8 training hours will be needed for a collections clerk, plus 20 hours more for her to meet with the programming staff initially and go over requirements for the new system. Once the system is installed, Mr. Anderson believes that the collections staff will virtually eliminate its record-keeping time, which will allow him to cut one staff person (who earns $25,000 per year, or $12.02 per hour) while still increasing the number of collection calls by 30%. The increase in calls should reduce the company's days of receivables from 47 to 42, which is a 10.6% decrease. The company's average total receivables balance is $4,525,000, and its borrowing rate of interest is 8%. Should the collections supervisor get his system?

The cost savings come from reducing the receivables balance as well as from reducing the number of staff members needed in the collections department. The analysis follows:

Cost of Collection Software

Programming time required	2,200 hours
Programming cost/hour	× $30
Total programming cost	$66,000
Collections time required	28 hours
Collections clerk cost/hour	× $12.02
Total collections cost	$337
Total cost of collection software	**$66,337**

Benefit of Installing Collection Software

Eliminate one clerk position	$25,000
Receivables balance	$4,525,000
Balance decrease with collection software	× 10.6%
Decrease in receivables balance	$479,650
Borrowing rate of interest	× 8%
Interest earned on reduced capital needs	$38,372
Total savings after installation	**$63,372**

With a one-time cost of $66,337 and savings of $63,372 in its first year of operation, the cost/benefit analysis indicates that the collection software will pay for itself in short order.

CALL CUSTOMERS PRIOR TO INVOICE DUE DATES

As the CFO of Fundamental Dynamics, Inc., you are concerned that many of your company's invoices are lost in the midst of your customers' approval processes. This problem is aggravated by the large size of the invoices for the capital equipment that the company sells. As a result, the collections staff is constantly faxing or emailing extra invoice copies to customers. The collections manager estimates that one-third of all customers lose their invoices (either temporarily or permanently) in the approval process and require new invoice copies. This results in a drastic lengthening of the days of receivables, from 42 days to 54 days (a 29% increase). The company's average total receivables balance is $3,150,000. The interest rate on the company's debt is 7%. The collections manager estimates that it will require an extra collections clerk to contact all customers in advance of their invoice due dates in order to figure out which one-third of the total customer base has lost its invoices. The collections clerk will cost $32,000. Should prior calls be implemented?

The cost savings are from reducing the days of receivables outstanding, whereas the labor cost associated with the extra phone calls offsets the savings. The analysis follows:

Cost of Calling Customers Prior to Invoice Due Dates	
Cost of collections clerk	**$32,000**
Benefit of Calling Customers Prior to Invoice Due Dates	
Average receivables balance	$3,150,000
Percent of balance due to lost invoices	× 29%
Amount of receivables due to lost invoices	$913,500
Interest rate on company debt	× 7%
Interest cost due to lost invoices	**$63,945**

Since the total cost is $32,000 and the savings are $63,945, the cost/benefit analysis indicates that customers should be called in advance. However, please note that the proportion of invoices lost in this example is only reasonable for capital equipment, because so many signatures are required for large invoices, resulting in invoices being lost as they are routed to collect signatures. An analysis for smaller invoices would yield a much lower percentage of lost invoices.

ELIMINATE ALL INVOICES

Mike Barrie, owner of Pan Pizza Equipment, Inc., has been contacted by several customers who would like to start paying the company based on their production. Pan Pizza produces the rollers that are used to move partially cooked pizzas through a baking oven. The customers will build the rollers into their baking machines, which require one week to build, and then pay Pan Pizza within two additional weeks. This means that payment will occur in three weeks instead of the usual 30 days (a reduction of 30%). The customers proposing this change make up 65% of Pan Pizza's total sales volume. Its average receivables balance is $1,560,000. The controller feels that there will be difficulty matching customer payments to any internal documents for control purposes, and proposes having some customized programming work done that will match payments received to the company's cumulative record of materials sent to each customer. The programming staff estimates that this project will require four months of programmer time (at an average salary of $70,000). The controller is also concerned that, without invoices, the company may have some difficulty proving to its auditors that it has booked the correct revenue each year. The controller decides to bring in the auditors for a review of the proposed transactions; this consulting work will cost $6,000. Also, the customized software will not automatically link to the company's packaged accounting software, so a one-day reconciliation must be manually performed each month by the assistant controller between the results of the customized software and the standard software. The assistant controller earns $75,000. Pan Pizza earns 5% on its investments. Should this new system be implemented?

The proposal has costs related to tracking the incoming payments and matching those payments to the company's sales. These costs are offset by greatly reduced accounts receivable balances. The analysis follows:

Cost of Eliminating Invoices	
Cost/year of programmer	$70,000
Four months' work	/3
Cost of programmer	$23,333
Cost of auditor review	$6,000
Cost/year of assistant controller	$75,000
Number of business days/year	/280
Cost/day of assistant controller	$268
Number of reconciliation days/year	× 12
Cost of reconciliation	$3,216
Total cost of eliminating invoices	**$32,549**
Benefit of Eliminating Invoices	
Average receivables balance	$1,560,000
Percentage reduction in receivables	× 30%
Dollar reduction in receivables	$468,000
Interest rate	× 5%
Total savings from reduced receivables	**$37,440**

With costs of $32,549 and savings of $37,440, the cost/benefit analysis indicates that the company should adopt a no-invoice system.

In summary, cost-benefit analyses can prove that the changes advocated in this chapter will be worthwhile to a company. However, the analyses are usually founded upon cost savings that do not relate to the speed of the transaction, since transaction speed is difficult to quantify. Thus, the accountant must frequently rely on other "incidental" savings when trying to justify changes whose primary objective is to streamline the accounting operation.

REPORTS

The changes discussed in this chapter require several new reports. For example, a pre-approved customer list is needed, as well as collection software reports. These reports contribute to the efficient, high-speed functioning of the accounting department.

A pre-approved customer list that includes the amount of total credit and available credit is useful for management review, since it highlights those customers with low credit and whose credit has been almost fully utilized. It should not be used each day by the staff to check against incoming purchase orders, since the report may become outdated quickly. It is better to have the order entry staff use on-line credit information in order to have the most up-to-date information. A pre-approved customer credit list is shown in Exhibit 2.8.

Exhibit 2.8   Pre-Approved Customer Credit List

Customer	Total Credit Granted	Total Credit Used	Unused Credit	Percentage of Unused Credit	Amt. of Over-due Invoices
Albatross Mining	$40,000	20,000	20,000	50%	2,000
Brass Refinishing	10,000	8,000	2,000	20	4,000
Cards R Us	5,000	0	5,000	100	0
Davidson Metals	12,000	1,000	11,000	92	0
End Run Printing	15,000	14,000	1,000	7	10,000
Gore Minerals	62,000	60,000	2,000	3	50,000
Highway Counters	33,000	13,000	20,000	61	4,500

Exhibit 2.9   Call List for Collections

Customer	Phone No.	Contact	Invoice No.	Amt. Due	Comments
Englehart Co.	719-112-6784	Evelyn	6742	$567.21	Faxed invoice
Fop & Mop. Corp.	508-231-4795	George	6751	330.33	Claims goods damaged
Grady's Ltd.	303-312-5678	Frank	6702	7,555.00	Need shipping trace
Honcho & Son	202-331-0932	Bill	6631	921.12	Claims goods damaged
Jobson & Clark	212-113-9074	Anna	6522	888.77	Emailed invoice
Killiwary Birds	308-222-3322	Mabel	6780	469.42	Paying on installment
Lanterns Plus	415-333-1111	Liz	6411	902.01	Referred to attorney

An appropriate use of computerization is the automation of the daily call list for the collections staff. The computer system can automatically update the overdue accounts list as money is received by the company, and also add overdue accounts to the list as they become overdue. An automated collections report is shown in Exhibit 2.9.

A good way to avoid overdue invoice payments is to call customers in advance to ensure that the invoice was received and that it has been approved and is ready for payment. If either of these has not occurred, then the collections staff can work on getting the proper information to the customer before the due date. A call report for upcoming invoices due is shown in Exhibit 2.10.

In summary, the sample reports shown in this section support bypassing steps in the normal sales transaction, reducing the work required to complete existing steps, and spotting problems as they occur, so that they can be fixed without delay. When used as part of the overall improvements mentioned in this chapter, these reports contribute to the increased speed of the sales cycle.

Exhibit 2.10   Call List for Upcoming Invoice Due Dates

Customer	Contact	Phone No.	Invoice No.	Invoice Amt.	Invoice Due Date
Englehart Co.	Evelyn Gregson	719-112-6784	6742	$567.21	5/31/09
Fop & Mop. Corp.	George Anders	508-231-4795	6751	330.33	5/30/09
Grady's Ltd.	Frank Horton	303-312-5678	6702	7,555.00	5/27/09
Honcho & Son	Bill Matthews	202-331-0932	6631	921.12	5/28/09
Jobson & Clark	Anna Davis	212-113-9074	6522	888.77	5/29/09
Killiwary Birds	Mabel Smith	308-222-3322	6780	469.42	5/31/09
Lanterns Plus	Liz Jones	415-333-1111	6411	902.01	5/30/09

METRICS

The suggestions for improvement in this chapter relate to improving the speed of the sales cycle and reducing the transaction processing work of the accounting staff. The metrics noted here are useful for spotting changes in these areas. They are as follows.

TIME FROM RECEIPT OF ORDER TO PRODUCTION SCHEDULING

This covers the time period in the sales transaction that applies to processing the incoming customer order. The best ways to improve this measure are to eliminate any move and wait times in the process and to bypass or automate as many other steps as possible. A final option is to perform functions in advance of the transaction (such as pre-qualifying customer credit).

TIME TO ISSUE AN INVOICE

This covers the time period from issuance of goods or completion of service to the issuance of an invoice. It can alternatively be lengthened to receipt by the customer of the invoice, to draw attention to the mail float. Examples of improvement techniques for this metric are to pre-issue invoices, assign specific responsibility for invoice issuance, and email delivery of invoices.

NUMBER OF DAYS' SALES OUTSTANDING

This covers the time period between when the product has been shipped but before the payment has been received, and tracks the speed of collection. The best ways to improve this metric are to rigorously reject the orders of those customers with bad credit records, review invoice approvals with customers prior to payment due dates, stream-line invoice formats, and accelerate the transmission of invoices to customers with email delivery. The days' sales outstanding formula is:

$$\text{Average no. of days' sales outstanding} = \frac{\text{Average receivables}}{\text{Annual sales on credit}} \times 365$$

For example,

$$\frac{\text{Average receivables}}{\text{Annual sales on credit}} = \frac{\$204,510,000/12}{\$122,220,000 - \$98,400} \times 365$$
$$= 51 \text{ days' sales out standing}$$

A well-managed collections operation should maintain a days' sales outstanding figure that is about one-third beyond the terms of sale. For example, if invoices are due in 30 days, an acceptable days' sales outstanding figure would be 40 days.

TIME FROM RECEIPT OF CASH TO UPDATING OF RECEIVABLES RECORDS

This covers the final phase of the sales and receivables transaction cy-cle. The best ways to improve this measure are to adopt an accelerated delivery of checks from the mail room to the accounting department, electronic access of lockbox records, and enforcing the immediate ap-plication of cash to outstanding receivable balances by the accounting staff.

NUMBER OF VALUE-ADDED STEPS IN THE TRANSACTION CYCLE AS A PERCENTAGE OF THE TOTAL NUMBER OF STEPS

This covers the entire transaction cycle and is an excellent measure of the efficiency of the entire process. Only about 10% of the steps in a typical transaction cycle are value-added. If the number of steps can be streamlined so that the ratio exceeds 50%, the process is extremely efficient. A ratio of 80% or better is world-class.

MEASUREMENTS TO TRACK REDUCTION OF THE WORK OF THE ACCOUNTING STAFF

The accounting manager should formulate and compile these costs periodically on a project basis to see if any changes in the costs are occurring. These costs are not ratios with specific formulas; the detail costs that make up each measure will vary by industry. However, the following can be used as a starting point for most businesses.

- *Cost to process an incoming order.* Divide these costs by the number of incoming orders to derive the cost per order:
  - Mail room labor to open mail and forward purchase orders
  - Labor to create a sales order based on the purchase order
  - Labor to conduct a credit check of the customer
  - Phone cost of contacting the customer about credit information
  - Labor cost of entering the order in the production system
  - Labor for filing of paperwork
  - Cost of physical storage
- *Cost per invoice issued.* Divide these costs by the number of invoices issued to derive the cost per order:
  - Labor to record shipping information in the shipping log
  - Labor to create an invoice based on the shipping information
  - Cost of the invoice paper and mailing materials

- ∘ Cost of postage
- ∘ Cost of mail room labor to mail the invoice
- *Cost of cash application.* Divide these costs by the number of payments received to derive the cost per receipt:
  - ∘ Mail room labor to open mail and forward checks
  - ∘ Labor to apply cash against receivables records
  - ∘ Labor to contact customers about application problems
  - ∘ Phone cost to contact customers
  - ∘ Labor to file invoice and payment information
  - ∘ Cost of physical storage
- *Cost of collection per invoice.* Divide these costs by the total number of invoices outstanding to derive the cost per invoice:
  - ∘ Labor cost to contact customers
  - ∘ Cost of phone calls to customers
  - ∘ Cost of overnight or fax transfer of information to customers
  - ∘ Labor for document filing
  - ∘ Cost of physical storage

INVOICES RETURNED FOR CORRECTION

Errors anywhere in the sales cycle tend to find their way into the final invoice sent to the customer. If any order or billing information was incorrect, the customer will return the invoice with a demand for changes. The accounting department should track the number of invoices requiring correction, as well as the reasons for each alteration. This is excellent source material for a sales cycle process review, since it tells the reviewer exactly where to look for problems.

In summary, most of the performance measures advocated in this section are not the standard ones included in most accounting systems; the accounting manager must decide if it is worthwhile to assign staff to track these new metrics. A reasonable solution is to track them on a sample basis periodically to see if they are worth the extra data collection effort.

SUMMARY

This chapter has highlighted a number of process best practices, technologies, reports, and measurement systems that, when used together, will accelerate the sales cycle. These changes involve reducing the move and wait times associated with the transaction cycle, reducing or forestalling the number of errors in the process, eliminating processing steps, and automating portions of the process. When all these changes are in place, the accounting staff will find that sales cycle transactions can be completed more quickly and with fewer errors.

Chapter 3

Cash

This chapter reviews a typical cash receipts system. It compares the typical system to a modified system that allows a company to process cash inflows more rapidly. The primary modification is the use of either lockboxes or lockbox truncation. In addition, the type of controls is shown to vary with the *method* of cash receipt (e.g., whether cash is received over-the-counter or through a lockbox) and the *form* of cash received (e.g., whether it is cash, check, or credit card charge information).

The modified system is then reviewed for control weaknesses and ways to improve the quality of its output. A detailed cost/benefit analysis is presented, which can be used as a model to determine if a company's cash systems should be converted to the modified system. The chapter concludes with a discussion of ways to measure cash processing performance.

CURRENT SYSTEM

Cash transactions have been burdened with more controls than any other type of transaction, because cash is easily removed from the company premises and liquidated. This section describes the broad array of cash receipt controls that are designed to reduce the incidence of unauthorized cash removal. This information is used later in the chapter to design a cash receipts system that involves fewer control points than a traditional system. Fewer controls translate into quicker transaction speed.

As outlined in Exhibit 3.1, the essential process flow is for check receipts to first be routed through the mailroom, where the mailroom staff creates a list of received checks that is used to ensure that all checks are appropriately accounted for once they reach the accounting department. The checks then go to the cashier, who manually enters receipt information into the cash receipts journal and prepares a bank deposit. The receivables clerk uses the check remittance advices or copies of the checks to record receipts against specific customer accounts in the accounts receivable ledger. Finally, the accounting manager completes the month-end bank reconciliation.

The controls noted in the flowchart are described at greater length as follows, in sequence from the top of the flowchart to the bottom:

1. *Mailroom prepares check pre-list.* As soon as the mail arrives, the mailroom staff should open all envelopes and prepare a list of checks that itemizes from whom checks were received and the dollar total on each check. It then copies this check pre-list, sending the original to the cashier and the copy to the accounts receivable clerk. A slight improvement in the control is to make an additional copy of the check pre-list and retain it in a locked cabinet in the mailroom, thereby providing evidence of initial receipt in case both the cashier and receivables clerk are in collusion and have destroyed their copies.

2. *Mailroom endorses checks "for deposit only".* By immediately stamping checks as "for deposit only" upon their arrival in the company, it becomes much more difficult for anyone in the accounting department to remove a check and cash it for their own use.

3. *Cashier matches check pre-list to cash receipts journal.* Once the cashier has recorded the amounts of all received checks in the cash

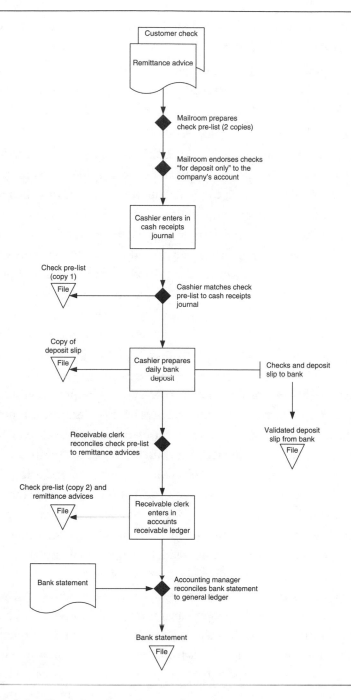

Exhibit 3.1   Check Receipts, Current System

receipts journal, this person should compare her entries to the check pre-list. By doing so, she can locate any errors in her entry.

4. *Receivable clerk reconciles check pre-list to remittance advices.* Though the first use of the check pre-list (by the cashier) was intended to reduce data entry errors, this second review is intended to prevent fraud by having a different person conduct the review.

5. *Accounting manager reconciles bank statement to general ledger.* Upon receipt of the monthly bank statement, the accounting manager should reconcile it to the accounting records; in a manual accounting environment, this calls for the use of subsidiary ledgers such as the cash receipts journal, since detailed records are not usually recorded in the general ledger, only batch totals. This control provides an independent review of both cash receipts and payable checks processed, and detects the removal of cash after it has been entered in the accounting system. This task should be performed by the accounting manager, rather than anyone in the cash handling or recording processes.

Traditional cash receipts processing systems require a great many paperwork transfers among employees. The move and wait times thus introduced greatly slow the cash receipts process. Also, every time a piece of paper is transferred, the potential exists for loss or misrepresentation. The value-added analysis shown in Exhibit 3.2 lists each step in an abbreviated process flow that only addresses the movement of checks from the mail room to the bank; even this shortened flow reveals the considerable amount of non-value-added time in the process. In the analysis, a value-added item is considered to be one that brings the cash transaction closer to conclusion.

Exhibit 3.3 provides a summary of the value-added analysis. It shows that only 7% of the steps bring the cash transaction closer to conclusion (depositing the cash at the bank); the remaining activities are related to moving paperwork from person to person or making file copies. Several steps exist only to cross-check the information that has been transferred between employees. In terms of time required, the

Exhibit 3.2 Cash Processing Value-Added Analysis

Step	Activity	Time Required (Minutes)	Type of Activity
1	Receive customer cash in mail room	1	Non-value-added
2	Wait – accumulate until all mail is opened	15	Wait
3	Prepare check pre-list	5	Non-value-added
4	File copy of check pre-list	1	Non-value-added
5	Hand-deliver checks and pre-list to cashier	5	Move
6	Checks wait in cashier's work queue	60	Wait
7	Cashier prepares daily bank deposit	2	Non-value-added
8	Cashier compares deposit slip to check pre-list	1	Non-value-added
9	Cashier files copy of deposit slip	1	Non-value-added
10	Cashier gives checks and deposit slip to bonded employee	1	Non-value-added
11	Bonded employee takes checks to bank	15	Move
12	Bonded employee deposits checks at bank	5	Value-added
13	Bonded employee brings receipts back to company	15	Move
14	Bonded employee leaves validated deposit slip in cashier mailbox	1	Non-value-added

value-added step can be concluded in five minutes, while the moving, waiting, and non-value-added steps take up over two hours.

The preceding process flow was designed primarily for the receipt of payments made by check. But what if the primary form of payment

Exhibit 3.3   Summary of Cash Processing Value-Added Analysis

Type of Activity	No. of Activities	Percentage Distribution	No. of Hours	Percentage Distribution
Value-added	1	7%	.08	4%
Wait	2	14	1.25	59
Move	3	22	.58	27
Non-value-added	8	57	.22	10
Total	14	100%	2.13	100%

is cash? Unlike checks, cash is completely untraceable, and thus is the preferred asset to steal. Thus, cash handling calls for tighter physical controls. The essential cash receipts process flow is for the initial cash receipt to be stored in a cash register, which is reconciled at the end of each shift; cash is removed at the end of each shift for deposit, after which the bank's validated deposit slip is reconciled to the company's original deposit slip. The flow is shown in Exhibit 3.4.

The process noted in the flowchart is as follows, in sequence from the top of the flowchart to the bottom:

- *Enter cash in cash register.* The primary role of the cash register is to record the amount of cash stored in it, either electronically or on a paper tape, while also providing a moderate level of security over the cash. If there is no cash register, as may be the case in very low-volume cash handling situations, at least use pre-numbered receipts to record the cash.

- *Give copy of receipt to customer.* When using a cash register, there is a risk that the cash register operator will remove cash and punch in a reduced cash receipt. To reduce this risk, always require cash register operators to give a copy of the receipt to the customer, since customers may review their receipts to ensure that the correct amount of cash was received. As an added inducement, many retail operations offer a free purchase to customers who do not receive a receipt.

- *Reconcile cash to cash register.* At the end of a cash handler's shift, a different person with no responsibility for cash handling should reconcile the cash in the cash register to the total of cash

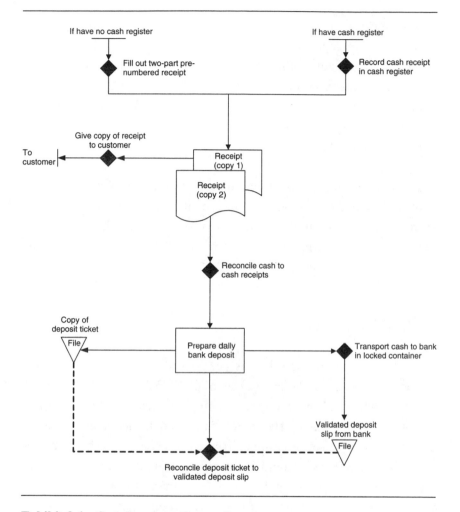

Exhibit 3.4   Cash Receipts, Current System

received as recorded on the register. Once completed, the person completing the reconciliation should sign and date it, so there is a record that a reconciliation indeed took place.

- *Transport cash in locked cash pouch.* To reduce the risk of unauthorized access to any cash being transported for deposit, always store it in a locked cash pouch. The most elaborate extension of this concept is to hire an armored truck to transport the cash, which is mandatory for larger quantities of cash.

- *Reconcile the validated deposit slip to the original bank deposit ticket.* Once deposited, the bank will issue a validated receipt for the cash. Someone other than the person who made the deposit should compare the original deposit ticket to the validated receipt and investigate any differences. This control is needed to ensure that the person making the deposit does not remove cash during delivery to the bank.

In summary, the series of actions required to complete a cash receipt transaction is complex, because of the high risk of theft or inappropriate use of funds. In the next section, we design several alternative cash receipt systems that require fewer steps to complete than the traditional system.

REVISED SYSTEM

While the foregoing controls may seem burdensome for a company that does not deal with large volumes of cash, they are needed in some industries where the volume of cash being handled is so large that the risk of loss is substantial. The trouble is that controllers tend to implement the complete set of cash controls even when the cash volume is so low that any losses would be minimal. In reality, the number of controls required varies significantly, from many for a casino to few for a company that transacts all business through barter exchange. In this section, we discuss how to speed up the flow of cash through alternative depositing techniques.

The single most important way to speed up the processing of cash is to implement a *lockbox system.* A lockbox is essentially a separate mailbox to which deposits are sent by customers. The company's bank opens all mail arriving at the lockbox, deposits all checks at once, copies the checks, and forwards all check copies and anything else contained in customer remittances to the company. The bank may also scan the checks and post them online for immediate viewing by the company. This approach has the advantage of accelerating

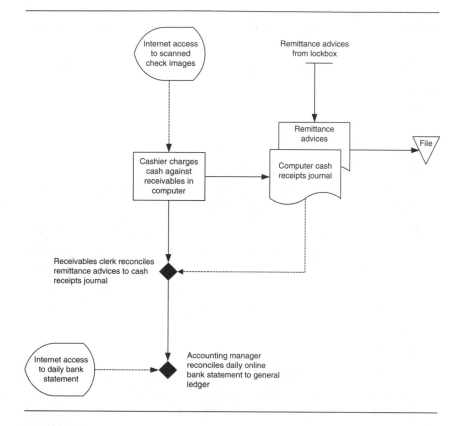

Exhibit 3.5 The System of Controls for Lockbox Receipts

the flow of cash into a company's bank account, since the lockbox system typically reduces the mail float customers enjoy by at least a day, while also eliminating all of the transaction-processing time that a company would also need during its internal cash-processing steps.

The number of controls needed in a lockbox environment is considerably reduced, since there is no cash on the corporate premises. The reduced process flow is shown in Exhibit 3.5. In this revised system, the cashier accesses check images over the Internet, or alternatively can obtain the same information from the remittance advices forwarded to the company by the bank. In either case, the cashier logs the receipts into the cash receipts journal, while the receivables clerk verifies that all receipts were correctly logged into the computer system.

The process steps noted in the flowchart are described at greater length as follows, in sequence from the top of the flowchart to the bottom:

- *Cashier charges cash against receivables.* The cashier accesses check images over the Internet, as posted there by the bank that operates the lockbox, and uses this information to record cash receipts in the company computer system.

- *Receivables clerk reconciles remittance advices to cash receipts journal.* On a daily basis, the receivables clerk prints the cash receipts journal for the date associated with remittance advices and check copies forwarded by the bank, and matches them to a printout of the cash receipts journal. This control ensures that all cash receipts are entered in the computer, and that receipts are charged to the correct customer accounts.

- *Accounting manager reconciles daily online bank statement to general ledger.* With online access to bank records, the accounting manager can conduct a daily review of the online bank statement, incrementally reconciling the bank account in the computer system. This control is an excellent detective method for quickly spotting any unusual transactions flowing through the cash account. A daily reconciliation is also useful for immediately recording any electronic payments and charging them to the correct customer account in a timely manner. This task should be performed by the accounting manager, rather than anyone directly involved in the cash recording process.

The primary difficulty with a lockbox is its implementation. When a company first asks its customers to start sending their payments to a lockbox, expect at least half of them to not switch on the first request. Instead, the accounting staff must wage a year-long campaign of letter writing, phone calls, and contacts with the management of each customer before the bulk of all payments are flowing to the lockbox.

Reasons for this are misroutings of the mailed request at the customer location, intransigent customer staff, or a customer who realizes that a lockbox will reduce its mail float.

A comparison of the controls required under a lockbox system to a basic cash receipts system makes it obvious that control systems can be greatly simplified if all cash is routed through a lockbox. However, customers are not always so accommodating in obediently sending their payments to a lockbox—there are always a few who persist in sending payments straight to the company. Consequently, unless arrangements are made with the mailroom staff to promptly mail these additional payments to the lockbox address, it will be necessary to retain the traditional cash handling controls.

If it is still necessary to continue processing checks in-house, the process flow can be improved through the use of *lockbox truncation*. Lockbox truncation is the process of converting a paper check into an electronic deposit. The basic process is to insert a check into a check reader, which scans the magnetic ink characters on the check into a vendor-supplied software package. The software sends this information to a third-party ACH processor, which typically clears payment in one or two days. Alternatively, the scanner creates an electronic image of the entire check, and transmits it directly to the bank. This approach removes from the typical check handling process the need for any daily bank deposit, though most other controls involved in the standard process flow noted earlier are still required. The modified process flow is shown in Exhibit 3.6.

The principal control addition in the exhibit is that the cashier should verify that the lockbox truncation report printed by the lockbox truncation software matches the checks just entered into the system, and then initials the report to indicate that this control has been completed. This is a simple error-correction control. Also, since the truncation report replaces the deposit slip, the truncation report is filed instead of the validated deposit slip. The original check can be retained, though there is no reason to do so after a few weeks, since the payment will have been processed by then.

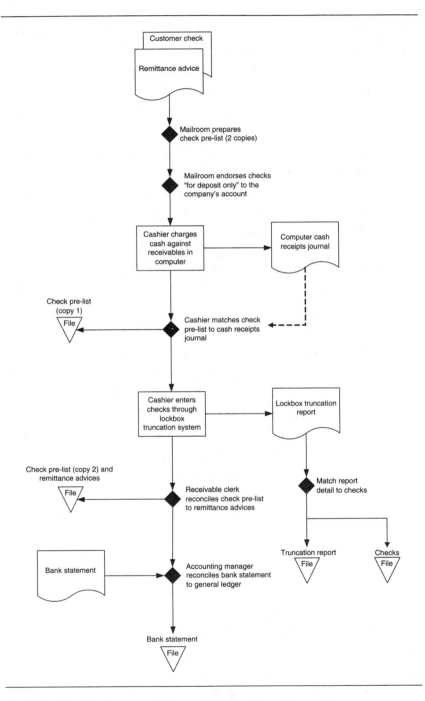

Exhibit 3.6   The System of Controls for Lockbox Truncation

In summary, it is difficult to shrink the blizzard of controls needed to process incoming checks and cash – subject to some possible alterations described in the next section. It is easier to shift the flow of cash to a lockbox, or at least a lockbox truncation system, which both require fewer controls.

CONTROL ISSUES

There are many control issues related to cash. For example, if the controller were to eliminate controls performed by the mail room employees, the risk of cash payments being intercepted by dishonest employees would increase. This section contains examples of what would happen if various cash controls were removed in order to increase the speed of cash transactions, and it suggests possible solutions to the resulting control problems. It also notes several key controls that should not be eliminated under any circumstances.

- *Mail room stops recording checks received.* Having the mail room staff record received checks is a control over the interception of checks. If they are recorded the moment they are received in the mail, the controller has a record that can be verified against the checks later deposited at the bank. If this control is not used, then for example, the cashier could remove checks and attempt to deposit them into a personal bank account. To avoid this issue, the controller can have checks go to a bank lockbox, thereby eliminating the check interception problem completely. Another variation is to accept a higher proportion of credit card payments or wire transfers, which also avoid the mail room.

- *Checks are not stamped "deposit to account."* This control is used to prevent an employee from cashing a check to a personal account. If the control is removed, the check can be deposited in an employee-owned account with a similar payee name. To avoid the issue, use a lockbox or accept credit card or electronic payments. In these cases, there is no check, so no deposit stamp is needed.

- *Consolidate cash handling tasks.* The controller can assign one person to handle all cash-related transactions, but then runs the risk of cash losses. For example, if one person receives, records, and deposits checks, the same person can divert checks for personal use, and never record their receipt in the accounting system. There are several ways to avoid this issue while still having one individual handle all cash receipts activities. First, continually shift the job among several employees. By doing so, any fraud would be short-lived. Second, enforce the taking of vacations. This keeps an embezzler from being on-site at all times, which is required to hide certain types of fraud. Third, ask the internal audit staff to regularly review the cash handling and recording function. Even if they find nothing, their ongoing presence will keep employees from attempting fraudulent activities. Typical internal audit reviews can include the comparison of cash register totals to cash counts, verifying cash sales against inventory records (to see if there is more inventory missing than would be indicated by cash sales), comparing mail room totals to deposit slips (to see if any checks were removed prior to the bank deposit), and tracing cash receipts to receivables (to see if incoming customer payments were diverted).

Some cash controls should *never* be avoided, since they leave too much risk of fraud. They are as follows:

- *Deposit all checks daily.* If checks are only deposited at the bank infrequently, the checks that are stored on site until the next deposit are subject to theft. This is also bad cash management, since the funds would otherwise be earning interest.
- *Always issue receipts.* Require all employees who receive cash (e.g., salespeople, tellers, store clerks) to issue receipts to customers for the cash received. A duplicate is also retained, so that the internal audit staff can reconcile the amount of cash received to the amount listed on the receipts. A variance may indicate that cash was withheld by the employee. Even when receipts are issued to customers, the person doing so may later write in a discount, which

he then withholds from the company. The best controls for this are reviews of receipt copies for evidence of tampering, and follow-up calls with customers to compare company sales records to their receipt records.

- *Review accounts converted to bad debts.* The controller should thoroughly review requests to write off bad debts. Though this may appear to be a control over receivables, it is also a control over cash; an employee can write off an account as a bad debt and then pocket the check when it arrives. When reviewing the bad debt request, the controller should not only review the probability of collection but also the number of bad debt requests submitted by each receivables clerk—a clerk who is pocketing customer checks may have submitted a disproportionate number of requests to convert receivables into bad debts.

- *Review requests for refunds.* Requests for payment refunds may not be coming from customers; they may be coming from clerks within the company. To guard against this, a request for refund should be on the customer's letterhead, which provides some guarantee regarding the source of the request. Of course, an enterprising employee could have created letterhead especially for the refund request, so a follow-up call to the customer is usually the best control over false refund requests.

- *Follow up on expected miscellaneous cash receipts.* Cash can be received for such miscellaneous items as the sale of company assets, insurance refunds, and legal settlements. The controller should track the dates and amounts of expected cash receipts and follow up on the cash settlements. Otherwise, these receipts may be pocketed. There is an additional danger with miscellaneous cash receipts, because they tend not to arrive in the mail room and be logged in with the checks mailed from regular customers. Instead, legal settlements may be delivered directly to the in-house counsel, asset sale payments may go to the office manager, and insurance refunds may be mailed to the in-house risk manager. Because of the method of delivery, they may not be listed on the mail room control sheet, and will thereby inadvertently skirt the company's cash control system.

In summary, there are a limited number of cases where cash controls can be reduced without seriously weakening a company's control structure. However, eliminating these controls will place a company at greater risk of monetary loss if it handles large volumes of cash or checks, and so their elimination is probably only advisable for low dollar-volume situations.

COST/BENEFIT ANALYSIS

This section contains an overview of how to a prepare cost/benefit analysis for implementing a lockbox. In the example, expected revenues and costs are as realistic as possible.

USE A LOCKBOX

The banker of Mr. Longfellow, president of Poetic Moments (a publisher of poetry books), has convinced him to use a lockbox for processing incoming checks. As a long-time controller, you are skeptical that the lockbox is worth the effort. Your research reveals that the bank charges $60 per month to keep a lockbox open, as well as a processing fee of 25 cents per check received. In a typical day, 80 checks would be received at the lockbox. An average daily deposit is $38,500, and the interest earned by Poetic Moments on its investments is 5%. The company would save one day of mail float by having checks sent to the lockbox. Internally, the mail room and accounting department staff are all hourly employees and earn an average of $17 per hour. It is company policy to send employees home when there is no work to do. The mail room staff person could be sent home a half-hour early if the incoming check pre-list no longer had to be prepared. Also, the cashier could save a half-hour per day by not having to prepare the deposit slip, and an accountant could save 15 minutes per day by not having to compare the pre-list to the deposit slip. Is it worthwhile to implement a lockbox?

The annual cost is $720 for the lockbox fee and $4,400 to process a year's worth of receipts (25 cents x 80 checks/day x 220 business days), resulting in a total annual cost of $5,120. The total annual savings are as follows:

Additional Interest Earned	
Average daily deposit	38,500
Interest/day	% (5%/365 days)
	× .014
Total interest/day	$5.39
No. of business days/year	× 220
Total interest/year	$1,186
Labor Savings	
Mail room	0.50 hours
Cashier	0.50 hours
Accountant	0.25 hours
Total savings/day	1.25 hours
No. of business days/year	× 220
Total savings/year	275 hours
Hourly rate	× $17.00
Total savings/year	$4,675

Thus, the total annual cost is $5,120, and total annual savings are $5,861, resulting in net savings of $741 per year. Based on the cost/benefit analysis, the lockbox should be installed.

The preceding cost/benefit example can be used to develop a real-life analysis, especially in terms of the line items used. However, the specific costs and unit quantities used in such an analysis should be derived from readers' experience, not from the example.

METRICS

The primary metric needed for the tracking of improvements to the cash process is cash flow speed. To measure it, use the customer's check date as a baseline and track how quickly this money clears the bank from the time it was sent to the company by the customer. This tracking can be difficult, since the check date and clearing date are not recorded in the databases of most accounting systems, and so would require the use of special database fields for tracking purposes.

SUMMARY

The maze of controls that surround the processing of cash are there for a good reason—to prevent the loss of cash by ineptitude or embezzlement. By and large, this chapter has not advocated the elimination of traditional controls. Instead, consider re-routing checks and cash away from the company, so that it never has to deal with them at all. Consequently, this chapter has focused on setting up a direct cash flow from customers to the bank via a lockbox. These techniques essentially replace high-control process flows with low-control flows that require less time and resources to complete.

Chapter 4

Inventory

In an accounting system, inventory is considered to be an asset. Anyone in the logistics department knows better—inventory is a liability that must be minimized at all times. If not, it will rapidly become obsolete, while excess amounts must be stored at considerable expense. Because of these costs, faster inventory transaction processing is a great boon to a company—it allows the organization an accelerated view of the exact status of all inventory items, so that it can reduce them.

This chapter examines the transactions most commonly used to acquire and track inventory, and also reviews a number of techniques for reducing both the cycle time and processing costs associated with inventory transactions.

CURRENT SYSTEM

Inventory transactions begin when the purchasing department orders parts that will eventually be delivered to the warehouse. This process is shown in Exhibit 4.1, along with key control points. The warehouse staff issues a pre-numbered purchase requisition when inventory levels run low. This is the primary authorization for the creation of a multi-part purchase order. One copy of the purchase order goes back

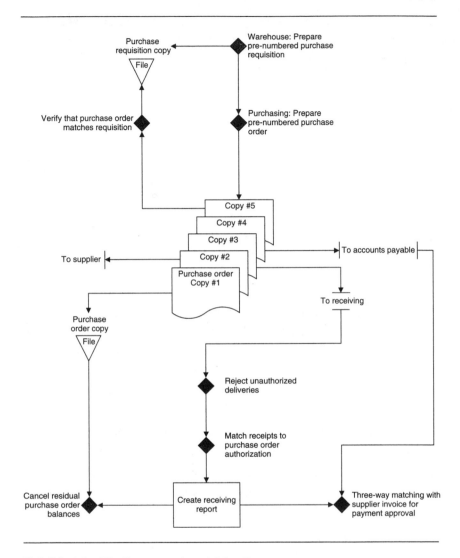

Exhibit 4.1 The Inventory Acquisition Process

to the warehouse, where the staff compares it to a copy of the purchase requisition to verify completeness; another copy goes to the supplier, while a third copy goes to the accounts payable department for eventual matching to the supplier invoice. A fourth copy is sent to the receiving department, where it is used to accept incoming deliveries, while a fifth copy is retained in the purchasing department. In short, various copies of the purchase order either trigger shipment from the supplier, or monitor its progress through the company's in-house systems.

Once the supplier delivers the new inventory, the warehouse must also inspect, tag, receive, and put away the items. Subsequent to storage, the warehouse must use a pick ticket to pick items from stock, and batch them into a kit for delivery to the production floor (or to the shipping department, if they are finished goods). The production department may then return excess parts to the warehouse or request any part for which there is a shortfall. During the production process, some parts may be scrapped or reworked, which requires additional documentation by the production staff. Once manufacturing is completed, the materials handlers shift finished goods to the warehouse, from which they are eventually shipped in response to customer orders. All of these steps are shown in Exhibit 4.2.

Thus, an inordinate number of transactions must be recorded to document the gradual advance of inventory items from suppliers, to the company, and eventually to the customer. Each transaction increases the probability of a record-keeping error, so the accounting or logistics department must spend an extraordinary amount of time tracking down errors. Also, the sheer volume of record-keeping may require more clerical time than actual manufacturing time to create a product.

The discussion so far has only addressed the physical flow of materials through a company. In addition, if a company uses the job costing system to assign costs to individual jobs, then a cost accountant must collect information about parts issued to or returned from a job, the direct labor hours used to manufacture it, and information about any items that were scrapped as part of the process. The accountant then summarizes this information, applies an overhead rate based on one or more rate bases (e.g., the total production labor cost, or the machine time used by the job), and issues a summary report when the job is completed. The cost accountant obtains much of this information from the production staff, which must have been recording this information during the production process. Because data collection by the production staff is considered secondary to the main business of

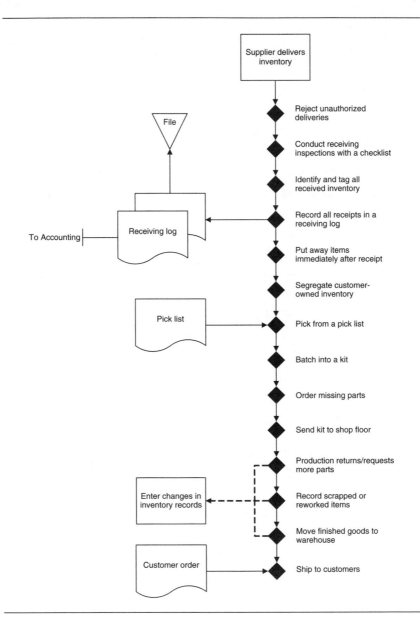

Exhibit 4.2   The Inventory Receiving Process

manufacturing products, the results tend to be slipshod, which means that the resulting job costing reports are also questionable.

Another time-consuming process is the physical inventory count. Typically run by an accounting manager, it requires a complete halt to the production process, while teams of clerks recruited from all over the company go to assigned counting areas, count items, and turn in count sheets. Another team of clerks compiles these sheets into an inventory summary report, which it compares to the book balance for each item. The count teams then recount any items for which there are large discrepancies, and adjust again as necessary.

The physical inventory counting process contains several serious flaws, which lead to the following errors:

- *Parts misidentification.* If non-warehouse clerks are used for counts, it will be truly surprising if they correctly identify parts with which they have no familiarity.
- *Unit miscounts.* A conscientious count team may still miss substantial inventory quantities, because those quantities are located elsewhere in the warehouse, or even on the shop floor. A less conscientious team will simply miscount the items that are directly in front of them.
- *Incorrect units of measure.* A counting team may assume one unit of measure when the inventory database uses a different one. For example, the count team may assume the unit of measure for a bag of parts is one bag, while the database assumes that each part within each bag is the proper unit of measure—the result is a gross undercount by the count team.
- *Missing tags.* Count teams may use thousands of inventory tags in a day, and it is a rare day when all of the tags are properly accounted for. More likely, a few will be stuck to other tags or simply lost. The result is inventory counts that are never tallied.
- *Transactions in transit.* Some receipts to or issuances from the warehouse that occurred prior to the start of the physical count may not have been recorded in the inventory database. The count team will adjust these balances to what they see during the physical count, and then warehouse staff will enter the transactions later, resulting in inaccurate inventory records.

In short, any time a company conducts a physical inventory count, there is a strong likelihood that the inventory record accuracy will decline as a direct result of the count.

Another set of problems arises from having multiple inventory databases, which must be combined to derive total inventory value. For example, a company may have a physical inventory, a separate perpetual inventory system (either a manual card file or a computer database) to track movements in and out of the warehouse, a database of inventory costs that is not linked to inventory quantities, and a general ledger that is not linked to any of the other information. This kind of scattered information causes the following book-to-physical adjustments:

- An item can be physically issued from the warehouse but not deleted from the physical inventory database. Alternatively, an item can be deleted from the database but still be in the warehouse.

- An item can be received into the warehouse but never added to the physical inventory database. Alternatively, an item can be added to the database but not be received into the warehouse.

- The cost of an item may have changed, but the new cost is not applied against the quantity in inventory. Alternatively, an item can be removed from the warehouse, and its old cost will still be reflected in the cost database.

- All of the preceding changes can occur, but never be recorded in the general ledger, or they may be recorded in the general ledger even though they never occurred.

The many steps in the previous flowcharts reveal a great many paperwork transfers between employees, which add time and the risk of lost or misinterpreted information to the inventory process. An inventory value-added analysis that begins with the receipt of parts at the warehouse is shown in Exhibit 4.3. In this analysis, a value-added item is considered to be one that brings the inventory transaction closer to conclusion. The table contains estimated wait times from kitted inventory and stored inventory parts. In actual situations, this wait time can

Exhibit 4.3 Inventory Value-Added Analysis

Step	Activity	Time Required (Minutes)	Type of Activity
[Parts arrive at receiving dock]			
1	Inspect parts for quality, quantity, and description	10	Non-value-added
2	Report problems to supervisor, who accepts or rejects the order	3	Non-value-added
[Shipment is accepted]			
3	Move problem items to a parts review area	2	Move
4	Store problem items in the parts review area	2	Non-value-added
5	Summarize quantity of remaining items on receiving form	5	Non-value-added
6	Move to copier	2	Move
7	Make two copies of completed receiving form	1	Non-value-added
8	Move to mailboxes	1	Move
9	Put one copy of receiving form in accounting department mailbox	1	Non-value-added
10	Move back to receiving department	2	Move
11	Move remainder of received shipment to warehouse	4	Move
12	Warehouse clerk inspects received shipment	5	Non-value-added
13	Warehouse clerk signs receiving document and takes possession of inventory	1	Non-value-added
14	Receiving clerk files signed copy of receiving document	1	Non-value-added
15	Warehouse clerk files copy of receiving document	1	Non-value-added
16	Warehouse clerk generates part number labels, and labels all new inventory	10	Non-value-added

(*Continued*)

Exhibit 4.3   Continued

Step	Activity	Time Required (Minutes)	Type of Activity
17	Warehouse clerk moves new inventory to warehouse storage area	2	Move
18	Warehouse clerk puts inventory in warehouse bins	12	Non-value-added
19	Warehouse clerk records location and quantity of all new inventory	2	Non-value-added
20	Warehouse clerk moves to computer terminal	1	Move
21	Warehouse clerk enters inventory item number, bin location, and quantity into computer	5	Non-value-added
22	Stored items wait until needed for production	[5 days]	Wait
[Production begins that requires inventory]			
23	Warehouse clerk receives pick list from production control department	1	Non-value-added
24	Warehouse clerk picks requested inventory items from shelf	20	Non-value-added
25	Warehouse clerk moves to computer terminal	1	Move
26	Warehouse clerk enters picking information into computer to take inventory items from perpetual inventory database and charge against production job	8	Non-value-added
27	Picked items wait until needed by production staff	[1 day]	Wait
28	Production staff compares pick list quantities to amount kitted by warehouse clerk	5	Non-value-added
29	Production staff signs pick list	1	Non-value-added
30	Warehouse clerk files signed pick list	1	Non-value-added

Exhibit 4.4 Summary of Inventory Value-Added Analysis

Type of Activity	No. of Activities	Percentage Distribution	No. of Hours	Percentage Distribution
Value-added	0	0%	0	0%
Wait	2	6	64.00	97
Move	8	27	0.25	0
Non-value-added	20	67	1.58	3
Total	30	100%	65.83	100%

stretch to many more days, depending on inventory turnover and how rapidly the production crew is processing current jobs.

Exhibit 4.4 provides a summary of the value-added analysis. It shows that none of the steps bring the inventory transaction closer to conclusion; all are related to moving paperwork from person to person, inspecting the inventory, or making file copies. When the only value-added step is getting the inventory to the production area, all these steps merely delay that move. In terms of time required, the process can be entirely eliminated, while the moving, waiting, and non-value-added steps take up over eight days. In short, the actions needed to conclude the transaction are a zero proportion of the total process.

In summary, many transactions are required to order an inventory item, receive it, move it from the receiving dock to the warehouse, and from there to the production area. In addition, information must be collected to track job costs while an item is being produced, and yet another set of transactions is needed to conduct a physical inventory at period-end. All these transactions combine to slow down the accounting and manufacturing processes. The next section discusses a revised inventory system that reduces many of these problems.

REVISED SYSTEM

From the accountant's perspective, improving the inventory area does not just involve streamlined inventory transactions, but also significant reductions in the overall amount of inventory on hand. The less inventory a company maintains, the fewer transactions the accounting staff needs to deal with.

The processes that impact inventory are spread throughout a company, and each one can be altered to either directly or indirectly reduce the accountant's job of monitoring inventory transactions. As detailed in this section, revised product designs can reduce inventory levels, as can modified costing systems, altered product forecasting, purchasing, receiving, storage, and so on. Very few of these changes are the responsibility of the accountant. Instead, the engineering, logistics, and production departments must be willing to alter their procedures in order to reduce inventory levels and streamline inventory-related transactions. Recommended changes fall into a number of categories, which are as follows.

PRODUCT DESIGN

- *Reduce the number of product options.* Design engineers like to offer a wide range of product options from which customers can choose. However, this requires that a large amount of inventory be kept in stock to deal with the full range of possible customer orders. The solution is to offer customers a greatly reduced set of product options. By doing so, the amount of required finished goods and sub-assemblies can be substantially reduced.

- *Eliminate redundant part numbers.* Part numbers may be redundant when a company uses multiple engineering teams to create a number of products at the same time, since each team may not be aware of part designations being made by their counterparts. It also occurs when a company switches to a new supplier, since the person assigning part numbers may not be aware of existing designations. This can be a significant problem, since the same part may be kept in inventory under multiple part numbers, which can noticeably increase the total inventory investment. The solution is to periodically make a concentrated effort to locate redundant parts, usually by working with cycle counters or scanning inventory records for similar descriptions.

- *Standardize parts.* The engineering department has likely created products without regard to the components incorporated into previous designs. The result is a broad range of fittings and fasteners in stock that are used in slightly varying sizes across a range of products, each of which must be kept in stock. A long-term solution is to standardize parts across multiple products. To do so, the engineering department should create an official "approved" list of standardized parts that can be used in new designs; in addition, entirely new parts require the approval of multiple managers. A useful tool in the identification of commonly used parts is the *matrix bill of materials*; this format displays the components of similar products in a side-by-side format, so that visual comparisons can be more easily made.

- *Avoid risky-procurement items.* Some product components can only be purchased from parts of the world having difficult legal, exchange rate, or political problems. For these items, the purchasing staff typically responds by stockpiling large quantities. To mitigate this problem, have the product design staff design alternative components into new designs, or at least reduce the amount of such parts needed.

- *Require approval of engineering change orders.* When the engineering department changes a product design, it does so with an engineering change order, which describes which old parts are being replaced with new parts, and any changes to associated production processes. When the change order goes into effect, this may maroon a considerable amount of old parts in the warehouse that are no longer needed. To avoid this excess inventory, have the logistics department review and approve all engineering change orders, with a particular emphasis on their reduction of old parts to the bare minimum, prior to switching to the new product configuration.

- *Delay the order penetration point.* Some products are specially configured for a specific customer. If a customer halts, delays, or cancels ordered items, the company may have difficulty disposing of the inventory. The solution is to maintain inventories at the highest possible sub-assembly level for as long as possible, and then add the last few product features when the customer order arrives. The engineering department can assist by designing products to be based on a common set of sub-assemblies that can be easily reconfigured into as many products as possible.

- *Require a purchase order for all new materials.* If all incoming items must have an authorizing purchase order, then they must, by definition, have an associated cost that will automatically be rolled up into the general ledger. This improvement only works if the purchase order software contains a mandatory requirement that a unit price be included for each item ordered.

- *Identify inactive inventory in the product master file.* When a company disposes of outmoded inventory, the computer system may automatically reorder the inventory, since the on-hand inventory balance has dropped to zero. To prevent this, have a procedure in place to always deactivate obsolete inventory in the product master file. This will prevent the system from ordering replacement stock.

PRODUCT COSTING

- *Use standard costs.* The accounting department may invest considerable resources in recording the costs of a specific job. Unless required by a customer contract, it is much simpler to use highly accurate bills of material to cost products. This eliminates much of the detailed cost accumulation work required for a comprehensive job cost analysis. This approach also keeps the production staff from having to submit time and cost reports. If there is a history of cost variances from standard, then the company may still need a variance tracking system, but only for the most significant variances.

- *Conduct a configuration audit.* A bill of materials is the basis for a large number of transactions, ranging from ordering to picking, as well as standard costing. If a bill is incorrect, it causes a ripple of problems throughout a company. To avoid this issue, schedule a periodic configuration audit for every bill of material as soon as it is released, as well as after an engineering change order. Under a configuration audit, an engineer or auditor who is familiar with the product pulls a completed product from the warehouse, disassembles it, and compares it to all engineering documents related to the product, including all authorized updates.

- *Eliminate purchase price variance reporting.* When a company purchases materials from multiple suppliers using short-term or spot pricing, purchase prices can vary wildly from the standard or historical purchase price. When this happens, the accounting staff reports the purchase price variance from a standard price, which management uses to castigate or reward the purchasing staff for their performance. Given the massive number of purchasing transactions in most companies, the accounting department can spend a great deal of time tracking this variance, while the analysis of it can be a waste of time. The solution is to have a long-term purchase agreement for each item, in which a price is fixed, and to then only report on changes from this fixed price.

- *Create a missing cost report.* Create a missing cost report that the accounting staff runs as part of its month-end closing process. The intent of this report is to present any inventory items for which there is on-hand inventory, but for which there is no cost in the database. The accounting staff uses this report to track down missing costs, thereby avoiding an under-reported inventory valuation.

- *Eliminate production scrap reporting.* Production workers must perform a carefully orchestrated production routine throughout the day. When a product must be reworked or scrapped, the cost accounting system requires them to stop what they are doing, log in the item on a form, and then return to what they were doing. This is highly inefficient. A better approach is to not bother employees with scrap reporting at all. Instead, focus their attention on fixing scrap-related problems at once, and have non-production people summarize all scrap for the entire production area at regular intervals.

- *Revise traditional cost accounting reports.* The traditional set of cost accounting reports are issued too late to be of use, involve information that managers cannot use, and are so voluminous that no one reads them. There are several ways to improve on the situation. First, cluster products and parts into related groups, with subtotals, so that the information is easier to analyze. Second, only issue a report if it can be done quickly enough to give actionable information to management. If not, do not issue it at all. Third, avoid overhead costs. In most cases, the allocation of overhead to a specific product results in an incorrectly low profit figure, and skews management decisions related to that product. Finally, and most important, only report on exceptions. If processes or products are operating within expected tolerances, then do not report them. Instead, focus management's attention on just those items requiring immediate attention.

PRODUCT FORECASTING

- *Link into customer planning systems.* Actively pursue direct linkages with the inventory planning systems of customers. By doing so, the company can eliminate a great deal of uncertainty from its planning process, and thereby avoids considerable amounts of excess inventory quantities.

- *Produce to order.* Inventory is at its greatest when a company builds to a forecast, and least when it only builds to specific orders, since the forecast is bound to be incorrect to some extent, resulting in unsold goods. Though not always possible (especially for seasonal sales), producing to order is the preferred way to reduce finished goods inventory.

- *Question the customer service level.* Many companies promise very high customer service levels by guaranteeing that inventory will be on-hand. This entails the use of very high inventory levels, so that every part is in stock at all times. Though a purely customer-oriented company will resist this method, it is still useful to periodically review the customer service level, to see if some inventory levels can be reduced in exchange for modest decreases in customer service.

INVENTORY ORDERING

- *Reduce supplier lead times.* A company goes to great lengths to reduce its internal lead times by a variety of just-in-time techniques, but accepts the supplier lead times without argument. These lead times are frequently not based on the supplier's actual production capabilities, but rather on policies. The result is long lead-times, which the company deals with by investing in excessively large safety stocks. To avoid this problem, have the purchasing staff include a shorter supplier lead-time requirement in its purchase orders.

- *Buy from suppliers close to the company.* When a company buys from suppliers located far from home, it typically orders in larger quantities in order to reduce the transportation cost per unit shipped. This tends to increase the amount of inventory on hand. Instead, consider buying only from local suppliers, so that smaller order sizes become cost-effective. The offsetting problem is that no high-quality suppliers may be located near the company.

- *Require smaller supplier deliveries.* If suppliers make more frequent deliveries of smaller quantities, there is less on-hand inventory to track. However, more frequent deliveries also entail much greater volumes of receiving transactions, so alternative procedures must be used to ensure that the receiving department is not buried by an avalanche of incoming orders. If a supplier objects to making a large number of small deliveries, then offset this requirement with a guarantee of long-term order volumes.

- *Eliminate departmental stocks.* When production managers receive more inventory than they immediately need, they have a tendency to squirrel it away on or near the shop floor, so they have a buffer available in case they run out of material. This creates extra inventory that is not recorded in any database, so costing records will be inaccurate. To avoid this problem, create a clean shop floor environment, where excess stocks are shifted back to the warehouse at once.

- *Require a purchase order.* If all received items must have an associated purchase order, then those incoming items without one can be segregated immediately for review. This procedure immediately isolates any customer-owned items, since they are not linked to purchase orders. By identifying and segregating customer-owned inventory immediately upon arrival, it is possible to avoid the considerable effort needed to identify such items when they have already been mixed into company-owned inventory.

RECEIVING

- *Reject unplanned receipts.* Unplanned and unidentified receipts can arrive at any time, requiring the receiving staff to set them aside for eventual identification, log-in, and disposition. This can take days and interfere with the orderly running of the receiving area, as well as build up undocumented inventory. These problems can be overcome through the rigorous rejection of all unplanned receipts. However, this path is extremely difficult to follow. The entire organization must understand that only authorized purchases that are initially routed through the purchasing department will be allowed at the receiving dock. This is a hard lesson to learn when an undocumented rush order arrives and is rejected.

- *Enter receipts into the computer at the receiving dock.* Some companies prefer to route all receiving documentation to the accounting department for data entry, on the grounds that the accounting staff is less likely to make data entry errors. Though there is some greater risk of entry errors at the receiving dock (generally avoided by having the best-qualified receiving person do all data entry), this is still the preferred location for doing so. It eliminates the time delay and risk of lost paperwork that arises when any documentation is shifted between departments.

- *Pay suppliers based on production records.* A company usually pays its suppliers based on receiving records, and this is one of the main reasons for the existence of a receiving department. However, if the basis for tracking received items shifts downstream from the warehouse to the production area, then there is less need for a receiving department. Under this new scenario, the company keeps essentially no inventory on hand, and instead works with suppliers to deliver parts on a just-in-time basis to the production area. If the company manufactures the product for which a supplier provides the components, then it is reasonable to assume that the supplier actually delivered the parts. The payment calculation is to determine the number of finished goods completed during the measurement period, multiply by the number of parts required to complete those finished goods, and add the number of scrapped parts. The company pays the supplier for this number of parts.

  This system requires extremely accurate bills of material, and scrap reporting. The company must also commit to assigning a single supplier to each part; otherwise, there would be confusion regarding which supplier provided a part. If these elements are available, then a company can reliably pay its suppliers the correct amounts without any receiving department at all.

  Suppliers will require considerable persuasion to adopt this technique, since they will complain that they will not be paid for items damaged during the production process, and for delivery overages. The company must answer this concern by tracking parts that are damaged during the production process, so that suppliers can receive full credit for the value of the delivery. Alternatively, if the supplier delivers too many parts, overages should be returned to the supplier—the whole point of this process is to reduce or eliminate inventories.

WAREHOUSE

- *Move fittings and fasteners to the production area.* Move all low-value fittings and fasteners out of the warehouse and into the production area. This reduces the scope of inventory counts, and all incoming and outgoing warehouse transactions, since as much as half of an inventory's total number of parts can be fittings and fasteners. This also reduces the number of bins in the warehouse devoted to low-value items, and eliminates the picking of these items from the warehouse shelves. Finally, consider hiring suppliers to both own and manage these items under a blanket purchase order agreement; this eliminates a large quantity of purchasing paperwork, and the associated receiving activities.

Not only is it not necessary to maintain inventory records for items shifted to the production floor, but the accounting staff should actively encourage its immediate expensing. Floor stock is generally not very expensive, and so there is only a modest initial impact on the financial statements when these items are charged to expense.

- *Use standard inventory containers.* If a company does not store or move inventory on pallets, then consider using standard containers instead. Depending on how such a container is set up, an inventory counter can ascertain its total quantity at a glance. A common storage and counting technique is to fill several of such containers with exactly the same quantity, and pick only from the most accessible container. By doing so, an inventory counter can easily determine the quantities in all other filled containers and manually count only the one partially-filled container in front. This concept can be taken too far. In many situations, it is easier to move, store, and count a partially-used pallet load without going to the considerable effort of shifting everything into standard containers. Its best application is for smaller parts that would otherwise be difficult to handle and count.

- *Eliminate obsolete inventory.* By continually examining the need for inventory and eliminating obsolete items, the warehouse can reduce the amount of inventory to be counted, stored, and tracked. Obsolescence tracking and disposition is best accomplished with a materials review board (MRB), which must periodically review inventory for obsolete items, authorize the elimination of selected items, and determine the most cost-effective way to do so. The best location method is the *where used report,* which itemizes the bills of material in which each inventory item is used. If the report reveals no associated bill of materials, then an inventory item can probably be eliminated.

An unusual aspect of eliminating obsolete inventory is that the inventory records for the remaining inventory items tend to become more *in*accurate. The reason is that obsolete inventory is the most reliably accurate part of any inventory audit; it never moves, and its quantity never changes. Once removed, the remaining inventory appears to turn over much more quickly, which leads to greater opportunities for transaction errors, and therefore a more inaccurate inventory.

- *Maintain a perpetual inventory system.* A physical inventory process can be avoided by setting up a perpetual inventory system instead, and ensuring that the accuracy of its records are kept at a high level. To do so, the warehouse manager must load all inventory records into an inventory tracking database, lock down the warehouse to keep away intruders, have experienced warehouse staff count small portions of the inventory every day (cycle counting), and continually investigate and correct the underlying problems causing any errors found. If the resulting system has a high record accuracy level, then it increases efficiency in several areas. First, there is no need to involve many employees in the period count, since there is no count. Also, production can continue at all times, which increases profits. Also, reduced parts shortages will yield fewer instances of expediting work through the production area, with the attendant disarray. Finally, there are fewer financial reporting surprises, since a perpetual system yields more predictable inventory balances.

Note that a perpetual inventory system should be implemented using the following sequence of steps. Not doing so can seriously prolong the time and expense needed to implement the system.

1. *Select and install inventory-tracking software.* Inventory-tracking software should record multiple locations for each inventory item,

update information in real time, and report inventory quantities by location (for counting purposes).

2. *Train the warehouse staff.* The warehouse staff should be well-versed in how to use the inventory-tracking software, especially the receipt, picking, and cycle count adjustment transactions.

3. *Revise the rack layout.* It is much easier to move racks prior to installing a perpetual inventory system, because no inventory locations must be changed in the database. Create aisles that are wide enough for forklift operation, and cluster small part racks together for easier parts picking. Also, if there is a small parts counter that issues parts to the production area, then cluster small parts racks near the counter. Designate separate racks for damaged goods and customer-owned inventory.

4. *Create rack locations.* Clearly label each rack location in accordance with a standard numbering system. As one moves down an aisle, the rack numbers should progress in ascending sequence, with the odd rack numbers on the left and the even numbers on the right. This layout allows an inventory picker to move down the center of the aisle, efficiently pulling items based on sequential location codes. Within each rack, there should be another designation for individual bins. For example, the warehouse location A-01-B means that an item is located in aisle A, rack 1, and bin B within rack 1.

5. *Lock the warehouse.* To stop the unrecorded removal of inventory from the warehouse, fence it in and lock all gates giving access to it. Only warehouse personnel should have access to the warehouse.

6. *Consolidate parts.* Group identical parts in the same bin, which reduces counting labor. Only allow an experienced warehouse person to do this, since anyone else will be more likely to mix parts together.

7. *Assign part numbers.* A mislabeled part is no better than a missing part, since the computer database will not show that it exists. Consequently, have only experienced warehouse staff label each part.

8. *Verify units of measure.* Have senior warehouse staff verify the units of measure for each part in the computer database.

9. *Pack the parts.* Pack parts into containers, seal the containers, and label them with the part numbers, units of measure and total quantity stored inside. Leave a few parts free for ready use. Only open containers when additional inventory is needed. These steps allow cycle counters to rapidly verify inventory balances.

10. *Count items.* Count items when there is little warehouse activity, such as on a weekend. This does not require multiple cross-checks, since subsequent cycle counts will locate errors. The main intent is to create a set of baseline records whose accuracy can be subsequently enhanced.

11. *Enter data into computer.* Only an experienced data entry person should enter the results of the initial count into the database. Once entered, have a second person cross-check the entered data against the original data for errors.

12. *Quick-check the data.* Print a report containing the initial count information, and scan it for errors. If all part numbers should have the same length, then look for ones that are too short or long. Review location codes for nonexistent racks. Look for units of measure that match the part being described. For example, is it logical to have a pint of steel in stock?

13. *Initiate cycle counts.* On a daily basis, have the warehouse staff print a portion of the inventory list, sorted by location, and count the items in that list. The sort by location is important, because a cycle counter can use the report to count within a concentrated area. Counters should look for inaccurate part numbers, units of measure, locations, and quantities. Counts can concentrate on high-value or high-use items, but the entire stock should be reviewed regularly. The most important part of this activity is to examine why mistakes occur, and alter procedures to reduce the risk that those mistakes will occur again.

14. *Initiate inventory audits.* A member of the accounting staff should audit the inventory once a week. Not only does this give rapid feedback to the staff regarding accuracy, but it also sends the message that inventory record accuracy is extremely important. For a material requirements planning system, 95% accuracy is the minimum acceptable target.

In addition, establish a tolerance level when calculating inventory accuracy. For example, if the inventory record for a box of screws shows a quantity of 100, but the actual count reveals a quantity of 105, the record is accurate if the tolerance is 5%, but inaccurate if the tolerance is 4%. The maximum tolerance should not exceed 5%, with a smaller tolerance for high-value or high-usage items.

The auditor should use as a source document an inventory list that does not reveal quantities. This keeps the auditor from conducting a "lazy" audit and simply checking off quantities on the audit report if counted quantities appear to be approximately correct.

15. *Post results.* Inventory accuracy is a team project, and the warehouse staff feels more involved if the audit results are posted against the results of previous audits. Also, it is cost-effective to motivate the warehouse staff toward greater inventory record accuracy by paying periodic bonuses that are based on attaining gradually higher levels of accuracy with tighter tolerances.

PRODUCTION

- *Use a pull production flow.* A just-in-time manufacturing system reduces or eliminates the storage and related tracking of inventory by keeping only enough materials in stock for the daily production schedule; all parts for production are delivered daily by suppliers. When a customer order arrives, the production facility orders just enough materials from suppliers (the "pull" production flow) to complete the order. Because the product is completed and shipped so quickly, there are almost no materials on site to be tracked, which eliminates many traditional inventory transactions. To make this system work, the company must certify each supplier in advance for predetermined quality and delivery criteria. By doing so, there is no need for a receiving inspection.

Supplier certification may take much longer than expected, because some suppliers will take a very long time to upgrade their system sufficiently to become certified; also, some suppliers may refuse to cooperate, requiring a search for an entirely new supplier.

- *Schedule smaller production batches.* Traditional cost accounting dictates that the setup costs associated with a production run can be reduced by spreading the cost over the largest possible number of units, which results in the lowest total cost per unit. The trouble is that this also yields massive inventory volumes and a greater likelihood of obsolete inventory. Accordingly, it is best to do the reverse and schedule smaller production batches, thereby reducing inventory quantities. This is especially effective when the production team emphasizes rapid equipment setup times, so that new production runs can be set-up with minimal expense.

- *Reduce container sizes.* Employees at production workstations generally fill an outbound container with products, and then move it to the next workstation once the container is filled. If the container is a large one, a considerable amount of work-in-process inventory can build up before it is time to move the container. If there are a number of workstations, and they all use the same size container, then partially-filled containers can represent a significant amount of inventory. To reduce in-process inventory, consider reducing the size of the containers. This concept can theoretically be taken down to containers of a single unit.

- *Use cellular manufacturing.* Traditional manufacturers keep all machines of the same type in the same part of the production area, so that jobs requiring the same type of processing can all be routed to the same area, and loaded into the first machine that becomes available. This tends to yield large inventory quantities piled up in front of low-capacity work centers, as well as more inventory moves between the various production centers. A better approach is to create manufacturing cells, where all of the equipment needed to create a product is located within a small area, and is serviced by a small group of experienced employees. This method yields extraordinarily low inventory volumes, since only one unit is produced at a time. Also, by slowing down the fastest machine in the cell to the speed of the slowest machine, it is impossible for a buildup of inventory to occur.

SHIPPING

- *Drop ship inventory.* The ultimate inventory goal is not to have any, and drop shipping is an excellent way to do so. When a company is buying finished goods from a supplier and then shipping it to a customer, it can simply notify the supplier to ship directly to the customer. By doing so, all traditional inventory transactions are replaced by a single shipment notification *to* the supplier, which is matched against a shipment notification *from* the supplier. However, many suppliers are unwilling to ship small quantities to individual customers, so this is not an option for many companies.

INVENTORY TRANSACTIONS

- *Reduce the number of stored data elements.* The typical company database contains a vast array of fields related to inventory, particularly in the item master file and bill of materials. When entering information into these records, one must tab through a large number of fields before arriving at the one requiring adjustment. This makes it possible to enter information into the wrong field. Further, a company may not use some of the information, so the time spent entering it was wasted. The solution is to limit the number of data elements being stored, by blocking selected fields from being displayed. This reduces tabbing and unnecessary data entry, and therefore the accuracy of the remaining information.

- *Eliminate transaction backlogs.* If there is a backlog of inventory transactions that have not been recorded in the inventory records, the reports used by cycle counters to test physical inventory will be incorrect. The cycle counters will certainly find differences between the inventory database and their physical counts, for which they will make adjusting entries in the computer system—which will *reduce* inventory accuracy once any unentered transactions are included in the inventory database. The obvious solution is to never have a transaction backlog, usually by allocating extra staff time to do them. Alternatively, have the warehouse staff use real-time, on-line data entry using wireless bar coded scanners, which keeps inventory records as up-to-date as possible.

- *Investigate negative inventory balances at once.* There is no logical reason why the inventory system should have negative inventory balances, so have the system flag them for immediate investigation. Investigation will reveal any number of possible reasons, for which the underlying problem can be investigated and fixed. Negative balances are an ideal indicator of a problem, and so should be investigated at once and with vigor.

For a more complete list of inventory-related improvements, consult the author's *Inventory Best Practices* book, which itemizes nearly 200 best practices.

In summary, there are dozens of ways to reduce inventory-related transactions by reducing the amount of inventory to which those transactions relate, as well as by switching to a "pull" system of materials flow. However, implementing most of the recommendations will initially require additional staff time. It is only after the implementation phase is complete that the accounting staff will experience reduced and simplified transaction volumes.

CONTROL ISSUES

A traditional inventory tracking system requires dozens of controls. For example, if inventory is not physically controlled within an inventory cage, employees may pilfer it. Similarly, removing controls over the recording of inventory receipts increases the risk of inventory being stolen before it even reaches the warehouse storage area. This section examines what would happen if various inventory controls were removed in order to increase the speed of inventory transactions, and suggests possible solutions to the resulting control problems.

- *Cancel the physical inventory count.* Replacing the physical inventory count with an appropriately administered perpetual inventory system actually *increases* the level of control. A complete physical count may take place only once a year, whereas a perpetual system requires the use of small, frequent cycle counts that, in total, result in a much greater number of item counts per year for the entire inventory. Consequently, it is much less likely that control problems will last long before being discovered.

- *Switch to a "pull" materials flow.* A pull system results in extraordinarily low inventory levels. If there is minimal inventory, it is difficult to see what control problems could remain. There are few items held in the warehouse whose existence could be improperly recorded, while any items on the shop floor are converted to finished goods so quickly that there is no point in tracking work-in-process inventory.

- *Move fasteners to the production area.* Shifting fittings and fasteners to the production area means that these items will no longer be listed in the inventory database, which would have made it easier to reorder at appropriate intervals. Also, by shifting them to the uncontrolled production area, there is a much greater likelihood that pilfering will increase. Though these are valid points, the transfer is *still* cost-effective. These items are so low-cost that the company can easily order them in excessively large quantities, accept a moderate amount of pilfering, and still experience a total expense less than if the items were stored in the warehouse.

- *Eliminate the warehouse.* The warehouse provides custody over materials, as well as a rigid control environment in which material quantities can be reliably tracked for costing and reordering purposes. The custodial function is always necessary if there are expensive goods, which cannot be ordered on a just-in-time basis, so the warehouse may still exist, but on a much smaller scale. Also, if there are high-quality bills of material and careful tracking of finished goods production, then this information can be used to derive reordering information and the cost of goods sold.

- *Stop receiving.* The receiving function is used to ensure that the correct quantity and type of items are received, and that their quality meets company standards. This department also logs receipts into the computer system. However, the engineering department can certify suppliers for quality, on-time shipping reliability and the correct number of parts delivered, so that the company no longer needs to conduct these tasks at the warehouse gate. Also, if goods are sent directly to the production floor, the company can use the finished goods count to determine how much was received through a *back-flushing* transaction.

- *Stop reviewing the period-end cutoff.* At the end of each accounting period, the accounting staff must ensure that all received items also have a cost in the computer system, so that the cost of goods sold is correctly calculated. Conversely, it must also ensure that inventory units are properly recorded to match any supplier billings. This review activity is usually considered to be a major accounting control. However, it is not necessary if there is a rigid policy in place of only receiving goods for which there is an authorizing purchase order, since the purchase order should always contain a price. Thus, it is impossible for items to be received without a matching cost automatically being recorded at the same time by the computer system.

Not all inventory controls should be eliminated. Several controls preserve the segregation of duties, affect the completeness of accounting paperwork, ensure that management reviews key transactions, or physically segregates inventory. The following controls are especially important:

- *Establish physical control over inventory.* Unless a company decides to eliminate inventory with a just-in-time "pull" system, the inventory must be controlled in a fenced-in, locked area. In addition, the warehouse manager must have total responsibility for inventory accuracy (as well as for any inventory shortages originating within the warehouse) and the recordation of all inventory transactions. Otherwise, it is too easy for a multitude of people to remove inventory from the warehouse without an offsetting entry in the inventory database. In short, if there is to be a warehouse, then it should have a rigidly-controlled environment.

- *Track obsolete inventory.* The proportion of obsolete inventory as a percentage of total inventory can rapidly spiral out of control, if there is not a continuing effort to track and dispose of it. The best control is a materials review board that meets regularly.

- *Audit the inventory.* It is very useful to have an internal audit person periodically conduct a book-to-physical comparison of the inventory. By doing so, there will be no surprises at the end of the fiscal year, when external auditors conduct the same review. Such an audit should verify that a) items physically in stock are recorded in the inventory database; b) items listed in the database are physically in stock; c) units of measure are correct; and d) large-dollar items are correct. The audit can also determine if there are significant changes on a trend line of the total number of parts, and of the inventory dollar value; a significant spike or drop in either one may indicate a larger problem.

- *Segregate the purchasing and inventory-keeping functions.* Continuing segregation of these functions will keep an employee from authorizing a purchase, and then diverting the incoming materials for personal use.

- *Account for the sequence of purchasing documents.* If a company relies on the purchase requisition or purchase order form as the primary tool to authorize a purchase, then the theft of either form is a serious control breach. These forms can be used by unauthorized staff to purchase materials in the name of the company, with shipment to some other location. However, this control is not a problem in a more automated environment, where the computer system itself

automatically issues orders based on forecasted inventory requirements; in this case, a company may not even use purchase order
documents.

- *Review parts costs by supplier.* Under a just-in-time manufacturing
  system, inventory is minimized, and consequently the risk of fraud
  is greatly reduced. However, to avoid collusion between buyers and
  suppliers (who no longer have to compete for purchase orders), the
  internal audit staff should periodically review supplier costs versus
  a sample of costs from other suppliers.

- *Use a receiving log.* For goods received through the warehouse,
  the receiving log remains a key control. If an item is not noted in
  the log, a company has no way of knowing if an item was received,
  and therefore cannot pay a supplier for the item without first
  receiving a proof-of-delivery document from the shipper (a
  time-consuming process that annoys suppliers). If the receiving
  department uses bar code or radio frequency identification (RFID)
  to log in items, then the computer system can automatically create
  a receiving log from this information.

- *Conduct closed job reviews.* One of the more useful control techniques is the after-the-fact review of closed jobs or production runs.
  Aided by a summary of costs, the management team should review
  each project and summarize the items needing improvement; this
  technique is useful for keeping management mistakes from occurring a second time.

- *Maintain an accurate bill of materials.* One of the most critical
  inventory-related records is the bill of materials, which lists the
  amounts and types of parts needed to build a product. If its constit-
  uent parts are highly accurate, it allows a company to predict the
  correct types and quantities of parts to purchase for an upcoming
  production run. A company can also use it to remove items from
  its perpetual inventory records once a product has been completed
  (called back-flushing). To verify the bill of materials information,
  cross-check additional items that are requested from stock during a
  job's production or returned to stock following production. These

transactions indicate if a bill of materials is incorrect. Also, ask the production and kitting staffs about repeatedly missing or duplicated parts. Finally, a review committee can systematically examine all bills of material for inaccurate quantities, part numbers, and units of measure.

- *Compile accurate scrap information.* Of particular concern when using a bill of materials to back-flush an inventory is the assumed scrap rate built into the bill of materials—if it varies greatly from the actual scrap rate, then actual inventory may vary substantially from the book balance. Also, if the bill of materials software only allows one scrap rate for everything listed in a bill, then back-flushing may result in incorrect inventory balances, because scrap rates can vary dramatically by individual component within a bill of materials.

In summary, controls surrounding the storage and movement of inventory can greatly increase the time required to complete inventory-related transactions. However, it is possible to selectively avoid some controls by adopting perpetual inventory tracking and just-in-time techniques that bypass the receiving and warehouse functions entirely. If the receiving and warehousing functions are to be maintained, then a minimum set of controls will still be necessary.

ERROR RATES

Inventory transactions require a great deal of manual labor to correct errors, so it is necessary to identify and root out the causes of errors on a continuing basis. The best ways to keep these errors from occurring are to reduce, eliminate, or simplify as many error-prone tasks as possible. The following bullet points identify the most common inventory-related errors, and how to mitigate them:

- *Incorrect bill of material changes.* There are two errors related to bills of material (BOMs): changes are made to the BOM database that are not reflected in actual usage, and changes are made to the product without updating the BOM database. The first error results in extra parts being purchased based on the incorrect BOM data, so that extra parts are ordered that are not needed. The second error results in items not being ordered, since the information is not listed in the BOM; this results in production delays, while parts are procured on an emergency basis. The best way to mitigate the first error is to limit access to the BOM database, so that only an authorized person with considerable experience can make changes. The BOM data entry screen can also incorporate limit checks, so that the software automatically rejects entries that do not fall within a pre-specified range.

Correcting the second error requires multiple, overlapping sources of information. Changes to BOMs can be reported by the engineer in charge of each product, the warehouse staff that receives excess materials back into the warehouse, by the warehouse staff that issues additional parts to the production area to complete a product, and by the production staff during their assembly of a product. By creating these multiple sources of information, it is less likely that changes will be made to a product that are not updated in the BOM database.

• *Incorrect units of measure.* Different departments prefer different units of measure. For example, the warehouse may prefer to count sheet metal in sheets, but the purchasing department may order it in pounds, and the engineering department may list it in square feet on product drawings. Thus, if an unauthorized person accesses an inventory record and changes the unit of measure, the quantity on hand can change drastically. For example, a 400-pound sheet of metal that costs $825 can become 400 sheets of metal that cost $330,000. To avoid this problem, limit access to the unit of measure field in the inventory database. The only person allowed access should be an authorized engineer or a supervisor. Another option is to have suppliers bar code the unit of measure on all shipments, so that the receiving staff can scan the correct unit of measure directly into the inventory database.

Some manufacturing software provides a table for listing multiple units of measure for each part. For example, sheet metal can be listed in sheets, pounds, or square feet, as shown in the following table. No matter which unit of measure is used, the transaction will still be accurate.

Description	Unit of Measure	Quantity
Sheet, stainless, 316L, 7 gauge	Lbs	378
	Sheet	1
	Sq Ft	48

- *Incorrect quantities.* A simple input error can result in an incorrect quantity in the inventory database. This error can be made by either the warehouse or receiving staffs whenever an inventory move occurs, or due to an incorrect count or delayed paperwork during a physical inventory count. One solution is to route all transactions through a warehouse clerk, who is presumably more experienced in recording the information. Another option is to use bar coded scanning of inventory transactions, so that more information is contained within a bar code, and therefore cannot be incorrectly entered in the computer system. Yet another solution is to use back-flushing, so that only a single entry at the end of the production process is needed to purge inventory from the system.

- *Incorrect warehouse locations.* When an inventory location code is incorrectly entered into the inventory database, the inventory is "lost," even though it is still in the warehouse. Cycle counting will eventually find the part, but this may take a long time. Another option is to label all bin locations with a bar code, which the materials handling staff then scans; this avoids any manual entry of information. Alternatively, a warehouse management system (WMS) can resolve this problem by directing the warehouse staff to store inventory in a specific location. A less efficient alternative is to reserve certain bins for specific inventory items, so that there is a default location listed in the inventory database; however, this can lead to a great deal of unused space in the warehouse. Finally, a computer limit check can prevent a data entry person from entering a non-existent location code.

- *Incorrect purchase order costs.* A part cost may be incorrectly input into the purchase order database, which is then used to cost inventory. If the cost is too low, the supplier will soon remedy the issue by calling to complain about not being paid enough. However, if the cost is too high, there is no built-in control to find it.

In summary, the best way to avoid inventory transaction errors is to have no transactions at all, which is best accomplished with an inventory "pull" flow system. Other alternatives are limiting access to database records, creating multiple review sources for key information, and using bar coding to avoid manual keypunching.

COST/BENEFIT ANALYSIS

This section demonstrates how to conduct cost/benefit analyses for implementing a perpetual inventory system, removing floor stock from the warehouse, eliminating the warehouse, stopping collection of actual work-in-process costs, and eliminating the receiving function. The expected revenues and expenses used in these examples will vary considerably from a company's actual situation, but the format used is a good framework for a realistic cost/benefit analysis. Examples are as follows.

IMPLEMENT A PERPETUAL INVENTORY SYSTEM

Bill Sweet, president of the Slow Times Syrup Company, wants to convert to a perpetual inventory system. He asks the controller, Mr. Honeycut, to analyze the costs and benefits related to the conversion. Mr. Honeycut finds that the company has had an average unexplained inventory write-down of $53,000 in each of the last five years after the year-end physical inventory was conducted. An extensive review of the bills of material indicates that they are accurate and have been accurate in the past. Also, the purchasing department calculates that it spends an extra $12,000 per year on rush freight charges to bring in materials that were supposedly in the warehouse, but could not be located. The warehouse must be enclosed for the perpetual inventory; it will cost $8,000 to fence in the warehouse area, plus $4,500 to install a computer and link it to the accounting software on the company network. In addition, two warehouse clerks must be hired at salaries of $25,000 each, plus a 10% overtime premium for the clerk working the second shift. Finally, two hourly employees must be assigned to the warehouse for three months to help with arranging the inventory, tagging it, and logging it into the computer; each of these employees is paid $14 per hour. Is it a good idea to install a perpetual inventory system?

The cost of installing a perpetual system versus the cost of *not* installing one is as follows:

Cost of Not Installing a Perpetual Inventory	
Cost of unexplained inventory write-down	$53,000
Cost of rush freight charges	+ 12,000
Total cost of non-perpetual system	$65,000
Cost of Installing a Perpetual System	
Cost of fencing in warehouse	$8,000
Cost to install a computer linked to network	+ 4,500
Cost/year of first-shift warehouse clerk	+ 25,000
Cost/year of second-shift warehouse clerk	+ 27,500
	$65,000
Cost/hour of employees for system setup	$14
Number of employees for setup	× 2
Hours in three-month setup period	× 520 hours
Total cost of employees for setup	$14,560
Total installation cost	$79,560

In short, the cost of not having a perpetual inventory is $65,000, whereas the cost of installing one is $79,560. However, much of the installation cost is a one-time occurrence, with only the $52,500 cost of two warehouse clerks continuing into future years. Thus, the new system becomes profitable after the initial installation period.

REMOVE FLOOR STOCK FROM THE WAREHOUSE

Mrs. Toadstool, the warehouse manager, has been ordered to cut her budget for the next fiscal year by 20%, which is a $10,000 cut. To do so, she has to reduce the hours of her cycle-counting employee. It appears that inventory accuracy will plummet if this happens. However, she realizes that there will be less inventory to count if she can move some of it directly to the production area. To justify this action, she collects a great deal of information. First, she prints out a list of inventory items, and identifies 500 parts out of a total inventory of 3,000 parts that are fittings or fasteners, and that cost 50 cents or less per unit. The total cost of these items is $14,000 out of a total inventory valuation of $662,000. When these items arrive, the warehouse staff uses a counting scale to bag them into clusters of 100 items per storage bag; this allows the cycle counter to count the parts more easily. The bagging process takes five minutes per receipt, and roughly 150 items are received per week. Parts are reordered based on a reordering report that is automatically printed every day and forwarded to the purchasing department. The cycle counter can count an average of 40 items per hour and counts the entire inventory once every two weeks. She discovers that a storage bin for each floor stock item (to be placed in the middle of the shop floor) can be purchased for $8 per bin, plus $5,000 for the rack (to be depreciated over ten years) and that a warehouse staff person can review the rack daily and take two hours to write down items that require reorders. A typical warehouse staff person is paid $12 per hour. If the bin is placed on the shop floor, pilferage is expected to be 10% per year. Is it worthwhile to move the floor stock out of the warehouse and into the production facility?

The cost of counting floor stock in the warehouse versus the cost of maintaining it on the shop floor is as follows:

Cost of Keeping Inventory in the Warehouse

Time/week bagging floor stock	12.50 hours
Weeks/year	× 52
Time/year bagging floor stock	650 hours
Labor cost/hour	× $12
Total inventory maintenance labor cost	$7,800
Number of floor stock items	500
Items counted per hour	40
Time to count floor stock every two weeks	12.5 hours
Number of times/year inventory is counted	× 26
Time/year to count floor stock	325 hours
Labor cost/hour	× $12
Total cycle count labor cost	$3,900

Cost of Moving Inventory to Shop Floor

Cost of floor stock	$14,000
Pilferage percentage	× 10%
Pilferage cost/year	$1,400
Time/week to review stock for reorder	2 hours
Weeks/year	× 52
Review hours/year	104 hours
Cost/hour	× $12
Cost/year to review stock for reorder	$1,248
Number of bins required	500
Cost/bin	× $8
Total bin cost	$4,000
Cost of rack for bins	+ 5,000
Total cost of rack and bins	$9,000
Number of years of depreciation	10
Depreciation cost/year	$900

In short, the cost of keeping floor stock in inventory is $11,700 per year, whereas the cost of keeping it on the shop floor is $3,548, which is a savings of $8,152 per year. This savings will allow cycle counting to continue for all other items remaining in the warehouse, while reducing the warehouse budget by nearly 20%.

ELIMINATE THE WAREHOUSING FUNCTION

The president of Custom Welded Products, a manufacturer of customized glove boxes for the nuclear industry, has reviewed the economic value added of each department and concluded that the warehouse adds no value to the finished product; therefore, it should be eliminated. As the controller, you explore the costs and benefits of this action. Fittings and fasteners can be moved to the shop floor, but 10% pilferage of the $10,000 of these items is expected. If no extra stocks are kept on hand, you estimate that $6,500 will be needed each year to order and ship in parts from suppliers on a rush basis, even if reasonably accurate bills of material are maintained. Also, an additional staff person must be hired in the engineering department to prepare more accurate bills of material, so that parts are ordered in exactly the right quantities. About 25% of the materials used in the products are 20% cheaper when ordered in bulk; if the warehouse is eliminated, they must be ordered in just the right quantities to meet production requirements. The company's annual materials cost is $1,500,000. The two warehouse staff people are paid $32,000 and $18,500. Also, $5,500 of annual depreciation will be eliminated if the warehouse racks are sold off. The resale value of the racks is $18,000. The raw materials inventory valuation is $1,250,000 and the company invests its short-term funds at an interest rate of 4%. Should Custom Welded Products eliminate its warehouse?

The benefit gained from eliminating the warehouse must be balanced against the costs associated with ordering smaller quantities, rush freight charges, pilferage, and the extra labor to maintain accurate bills of material. The analysis follows:

Benefit of Eliminating Warehouse

Eliminate warehouse salaries	$50,500
Sell racks	+ 18,000
Reduction in depreciation	+ 5,500
	$74,000
Cost of inventory	$1,250,000
Cost of floor stock retained	− 10,000
Net savings on inventory	$1,240,000
Interest rate on investment	× 4%
Net earnings from additional working capital	$49,600

Cost of Eliminating Warehouse

Cost of items moved to shop floor	$10,000
Expected pilferage	× 10%
Total expected pilferage cost	$1,000
Cost of rush shipments	$6,500
Cost of bills of material employee	$32,000
Cost/year for materials	$1,500,000
Materials cheaper in bulk	× 25%
Cost of bulk materials	$375,000
Small-order-quantity surcharge	× 20%
Cost of small-order-quantity surcharge	$75,000

The total cost of eliminating the warehouse is $114,500, and the total benefit is $123,600. Thus, it is cost-beneficial to eliminate the warehouse.

STOP COLLECTING ACTUAL WORK-IN-PROCESS COSTS

Mr. Smith, the controller of Steady State Systems, has just met with Ms. Jones, the controller of Amalgamated Products Unlimited, and discovered that her accounting department is two-thirds the size of Mr. Smith's, even though both companies are similar in size and function. Upon further inquiry, Mr. Smith finds that the other controller does not report on actual work-in-process costs, relying instead on bills of material generated by the engineering department and updated with selected information gathered from the shop floor. Mr. Smith is determined to try this as well, but must convince the top managers of Steady State Systems that they will still receive high-quality cost information despite the associated reduction in personnel costs. After considerable investigation, Mr. Smith finds that Steady State's bills of material currently show costs that vary from actual costs by an average of 15%. In order to report information that will not lead to poor decision making, Mr. Smith must reduce this variance to 5%. To do so, an engineer must be hired to continually update bills of material information; this person's salary will be $55,000. Also, the warehouse staff must report on items returned to the warehouse from the shop floor, as well as on extra parts issued to the shop floor (which reveals excesses or shortages on the bills of material). This will require a half-hour of warehouse time every day; the average warehouse worker earns $28,000 per year. In addition, the results of the existing scrap reporting system must be channeled to the engineer, who will incorporate this information into the bills of material; there is no cost associated with this step. Finally, the bill of materials engineer must meet with the production supervisors every month to review the labor rates shown on the bills of material and adjust them as necessary; the cost of supervisory time for this process is $8,000 per year. If the existing work-in-process accounting system is eliminated in favor of costs based on bills of material, then Mr. Smith can eliminate a cost accountant and an accounting clerk from the payroll. The cost accountant earns $48,000 per year, and the clerk earns $32,000 per year. Based on this costing information, should Steady State Systems switch to bill of materials costing? The analysis follows:

Cost of Bill of Materials Costing

Cost/year of bill of materials engineer	$55,000
Cost of review time with supervisors	+ 8,000
	$63,000
Average warehouse pay/hour	$13.46

(Continued)

Time/day to update bills of material	× 0.5 hour
Cost/day to update bills of material	$6.73
Number of business days/year	× 260
Cost/year to update bills of material	$1,750
Total cost for bill of material costing	**$64,750**
Cost of Actual Work-in-Process Costing	
Cost/year of cost accountant	$48,000
Cost/year of accounting clerk	+ 32,000
Total pay of work-in-process reporting staff	**$80,000**

In short, the cost of starting up a bill of materials costing system is $64,750, versus a cost of $80,000 if the current actual work-in-process reporting system is maintained. Based on this information, the bill of materials costing system should be implemented.

ELIMINATE THE RECEIVING FUNCTION

The CFO of Fonicka Cameras, Inc., a manufacturer of high-quality cameras, wants the controller to cut costs in the receiving area. The controller finds that only one of the two receiving employees will be needed if the company certifies its suppliers and has them deliver products directly to the camera production line. The company will save $25,000 by eliminating the receiving position, plus an additional $12,500 by reducing the other position to part-time. However, it must still maintain a receiving computer workstation to handle deliveries by smaller suppliers who are not certified. Also, two warehouse positions, each costing $23,000 per year, will be eliminated, since no items will be processed through the warehouse. The cost to monitor and certify suppliers is $60,000 in the first year, which pays for the part-time labor of an engineer and a purchasing agent, plus their travel costs to visit suppliers. This certification cost is expected to go down to $30,000 after the first year. In addition, the cost of purchased parts is expected to be unchanged when more frequent deliveries are enforced. The cost would have been higher, but all parts will now be single-sourced, so suppliers will absorb the delivery cost in exchange for higher purchased volumes. Finally, payments to suppliers will be made based on completed production volumes; the number of parts used in each completed item is based on a bill of materials. The number of parts in each bill must be carefully reviewed to ensure proper payment, and this requires the half-time labor of a $35,000 junior engineer. Is it worthwhile to eliminate the receiving function? The analysis follows:

Cost of Eliminating the Receiving Function	
First-year cost to certify suppliers	$60,000
Cost to review bills of material	+ 17,500
Total cost of eliminating the receiving function	$77,500
Cost of Retaining the Receiving Function	
Cost of retained warehouse positions	$46,000
Cost of retained receiving positions	+ 37,500
Total cost of retaining receiving function	$83,500

The receiving function has several warehouse positions tied to it, so they are also eliminated if the receiving jobs are no longer needed. Thus, the total cost associated with keeping the receiving function is higher than the cost of just the receiving payroll. Consequently, Fonicka Cameras should eliminate its receiving function in order to reduce total costs.

The cost/benefit examples in this section can be used as a basis for actual analyses. The examples are linked, so that the costs associated with eliminating part of the receiving function can also be used in justifying the elimination of the warehouse. Thus, when constructing an actual cost/benefit analysis, a better case can be made for change if several of these changes are combined into one cost justification proposal. Also, it would be more accurate to present costs and benefits over a longer time frame and to include a net present value analysis of the longer-term stream of cash inflows and outflows. Cost justifications tend to be more convincing over longer periods, since one-time project setup costs can be offset by labor savings that continue to pile up over future periods.

REPORTS

The reports needed for a revised inventory system all require approximately the same information but are sorted differently, and may require minor changes to data elements depending on the purpose of the report. These reports are used to maintain a perpetual inventory system that permits the accountant to quickly close each accounting period and track the accuracy and potential obsolescence of the inventory. The information on these reports should include each item's inventory location, item number, description, unit of measure (U/M), and quantity. Some reports may also require unit costs and extended costs. By altering the presentation of these reports according to different sort criteria and layouts, this information can be used for cycle counting, inventory audits, checks for mispriced items and incorrect units of measure, and checks for obsolete items. In this section, a sample report is shown for each application.

INVENTORY CYCLE COUNTING REPORT

This report is used by the warehouse staff to count blocks of inventory. It is sorted by inventory location. The report may print a blank line in place of the inventory quantity, thereby forcing the inventory counter to manually fill in the inventory quantity, rather than conducting a quick comparison of the quantity listed on the report to the amount in the bin. A manual count tends to be more accurate than a quick comparison. An inventory cycle counting report is shown in Exhibit 4.5.

INVENTORY AUDIT REPORT

This report is used by the internal audit staff to determine the accuracy of the total inventory. It is identical to the inventory cycle counting report, except that inventory items are selected at random, so that the

Exhibit 4.5   Inventory Cycle Counting Report

The Henderson Grape Drink Company
Cycle Counting Report
Date: __/__/__

Location	Item No.	Description	U/M	Qty
A-10-C	Q1458	Switch, 120V, 20A	EA	___
A-10-C	U1010	Bolt, Zinc, 3 × 1/4	EA	___
A-10-C	M1458	Screw, S/S, 2 × 3/8	EA	___
A-10-C	M1444	Weld Stud, 3/8 × 3/8	EA	___
A-10-D	C1515	Flat Bar, 304, 1 × 3	FT	___
A-10-D	C1342	Square Bar, 316, 2″	FT	___
A-10-D	C1218	Round Bar, 305, 1-1/2″	FT	___
A-10-D	C1110	Weld Pipe, 316, 7″	FT	___
A-10-E	A2700	Sheet Metal, 316, 7 GA	LB	___
A-10-E	A2710	Sheet Metal, 304, 9 GA	LB	___

auditor can conduct a broad-based review of the inventory. If the cycle counting report were used for this purpose, the auditor would only review a small portion of the inventory at one time, which may have just been cycle counted, and which may therefore not give an accurate finding regarding the overall accuracy of the inventory. An inventory audit report is shown in Exhibit 4.6.

INVENTORY VALUATION REPORT

This report is used by the accounting staff to review the valuation of inventory items. It includes all the information on the cycle counting and audit reports, as well as the unit cost and extended cost of each item. When it is sorted in descending order of extended cost, an accountant can review the most expensive items for accuracy. Usually, a quick comparison of the extended cost to the part description will suffice to reveal any items that have incorrect extended costs. One of the primary reasons for an inaccurate extended cost is an inaccurate unit of measure, so it is important that the units of measure be included in the report. Also the accounting staff should occasionally review the

Exhibit 4.6 Inventory Audit Report

The Henderson Grape Drink Company
Inventory Audit Report
Date: __/__/__

Location	Item No.	Description	U/M	Qty
A-08-C	M1471	Screw, Tap, $3 \times 1/4$	EA	___
A-12-D	M1100	Bolt, Hex Head, $2 \times 1/8$	EA	___
A-17-A	M0900	Bolt, Carriage, $4\text{-}1/2 \times 1/2$	EA	___
B-03-B	R0100	Fire Shield, $8'' \times 12\text{-}1/2''$	EA	___
B-05-E	R1109	Fire Shield, $4\text{-}1/2'' \times 12\text{-}3/8''$	EA	___
B-12-B	R7621	Fire Shield, $6\text{-}1/4'' \times 10\text{-}1/2''$	EA	___
C-07-E	C6721	Flat Bar, 316L, $1/4'' \times 4''$	FT	___
D-04-A	C0991	Square Bar, 304L, $4''$	FT	___
D-10-C	C8712	Square Bar, 316, $2''$	FT	___
E-08-A	A7720	Rapid Transfer Port, $8''$	FT	___

less expensive items to see if the reverse has occurred – that very expensive items are being cost at excessively low valuations. An inventory valuation report is shown in Exhibit 4.7.

INVENTORY USAGE REPORT

This report is used by the logistics department and the accounting staff to pinpoint low-usage items for deletion. It shows all the information in the inventory valuation report, plus the last date of use. The report is sorted by date, with the oldest dates first. The items with excessively old last-use dates are possibly obsolete. Items thus noted are subjected to a review by the materials review board to determine if they can be used or if they should be dispositioned. The accounting staff can use this report as a source of information for the obsolete inventory reserve, since the extended cost of possibly obsolete items is listed here and can be summarized to derive a total obsolete inventory figure. An inventory usage report is shown in Exhibit 4.8.

Exhibit 4.7   Inventory Valuation Report

The Henderson Grape Drink Company
Inventory Valuation Report
Date: __/__/__

Location	Item No.	Description	U/M	Qty	Cost ($)	Total Cost ($)
C-04-B	C1180	Square Bar, 316L, 4″	FT	150	62	9,300
E-08-A	A7720	Rapid Transfer Port, 8″	FT	9	972	8,748
D-02-D	U1010	Isolator Shell, 4′ × 8′	EA	5	995	4,975
B-03-C	R0100	Fire Shield, 8″ × 12-1/2″	EA	13	182	2,366
E-07-D	W0009	Switch, 120V, 20A	EA	29	70	2,030
D-04-A	C0991	Square Bar, 304L, 4″	FT	11	60	660
A-03-B	D3425	Flat Bar, 304L, 1″ × 4″	FT	42	14	588
F-12-C	J1482	Pipe, Alum, 8″	FT	13	42	546
C-10-C	Q5478	Silicone, White	EA	430	1	430
G-03-A	M1457	Screw, Hex Head, 1 × 0.5	EA	400	0.5	200

In summary, similar report formats can be used for a multitude of purposes: inventory cycle counting, auditing, obsolescence reviews, and cost extension reviews. When used together, these reports allow the accounting staff to maintain an accurate perpetual inventory, which in turn eliminates the need for physical inventory counts.

METRICS

Several measures give some indication of a company's inventory-related performance. However, the reasons why inventory metrics fluctuate must be understood, for bad management practices may underlie an otherwise reasonable measurement. Several useful metrics follow.

INVENTORY TURNOVER

The most widely-used measure of inventory performance is turnover. It measures the manufacturing system's efficiency in using inventory and is

Exhibit 4.8   Inventory Usage Report

The Henderson Grape Drink Company
Inventory Usage Report
Date: __/__/__

Location	Item No.	Description	U/M	Qty	Cost ($)	Total Cost ($)	Last Used
C-04-B	C1180	Square Bar, 316L, 4″	FT	150	62	9,300	01/09/08
A-12-D	M1100	Bolt, Hex Head, 2 × 1/8	EA	27	1	27	04/07/08
A-10-D	C1218	Round Bar, 305, 1-1/2″	FT	58	6	348	05/27/08
A-10-C	M1444	Weld Stud, 3/8 × 3/8	EA	992	2	1,984	06/12/08
D-02-D	U1010	Isolator Shell, 4′ × 8′	EA	5	995	4,975	07/03/08
D-10-C	C8712	Square Bar, 316, 2″	FT	117	25	2,925	12/28/08
E-07-D	W0009	Switch, 120V, 20A	EA	29	70	2,030	03/30/09
A-03-B	D3425	Flat Bar, 304L, 1″ × 4″	FT	42	14	588	04/13/09
C-10-C	Q5478	Silicone, White	EA	430	1	430	04/14/09
A-10-E	A2700	Sheet Metal, 316, 7 GA	LB	782	9	7,038	06/06/09

derived by dividing the usage factor by the average inventory. For example, the turnover of various inventories would be determined as follows:

- *Finished goods.* Cost of goods sold/average inventory of finished goods
- *Work-in-process.* Cost of goods completed/average work-in-process inventory
- *Raw materials.* Materials placed in process/average raw materials inventory
- *Supplies.* Cost of supplies used/average supply inventory

The result is the number of turns, usually measured in turns per year. Turnover statistics must be analyzed with caution, for different causes can underlie the same result. A slow turnover can indicate over-investment in inventories, obsolete stock, or declining sales. However, it may simply mean that a company is stocking up for a large custom job with parts that have long lead times. A very high turnover can indicate improved utilization through conversion to a just-in-time or material requirements planning system, or it may be caused by keeping excessively small stocks on hand, resulting in lost sales or increased costs due to fractional buying. Many industry organizations publish their average turnover rates, so benchmark information may be available.

The purpose of business is turning a profit, not turning inventory. Evaluating a company's performance based on just the turnover metric is not wise without more detailed information. If turnover is used to evaluate the performance of a new manufacturing system, then it is useful. If it is used to compare performance between accounting periods, it is useful as an indicator of underlying problems or improvements that must be researched further to determine the exact causes of any changes in the metric.

INVENTORY ACCURACY

The accuracy of the inventory database is measured as the total number of errors discovered, divided by the total number of inventory items reviewed. For purposes of this calculation, there can only be one error per inventory item (otherwise there might be a negative accuracy percentage!).

Errors in inventory can be caused by incorrect quantities, locations, and units of measure for an inventory item. A high inventory accuracy number means that the information in the inventory database can be relied on to yield accurate information about available inventory, which can be used for purchasing and production control decisions. The inventory accuracy metric can be skewed if the items are counted in the most recently cycle-counted area of the inventory; instead, the count should randomly review the entire inventory.

BILL OF MATERIALS ACCURACY

The accuracy of the bill of materials database is measured as the total number of errors discovered, divided by the total number of bills of material reviewed. For purposes of this calculation, there can be only one error per bill of materials.

Errors in bills of material can be caused by incorrect part quantities or units of measure, and the level of the bill at which a part is included. A high bill of materials accuracy number means that the information in the database can be relied upon to yield accurate information about the parts content of items to be produced, which can be used for purchasing and production control decisions.

INVENTORY SUPPORT COST

This metric is used to determine how much a company is spending to maintain its inventory. An excessively high ratio may trigger a cutback in warehouse-related expenses or a decision to adopt a just-in-time or manufacturing requirements planning system, both of which can reduce inventory levels and support costs. The calculation follows:

$$
\begin{array}{l}
 \text{Cost of warehouse salaries and related benefits} \\
+ \text{Cost of warehouse depreciation} \\
+ \text{Cost of interest on money invested in inventory} \\
+ \text{Cost to rent warehouse space} \\
+ \text{Cost of insurance on inventory} \\
\underline{+ \text{Cost of obsolete inventory}} \\
= \text{Cost of total inventory}
\end{array}
$$

PERCENTAGE OF NEW PARTS USED IN NEW PRODUCTS

This metric is used to encourage the engineering staff to re-use existing parts in new products as much as possible. The calculation is to divide the number of new parts in a bill of materials by the total number of parts in the bill of materials. Some companies do not include fittings and fasteners in their bills of material, since they keep large quantities of these items on hand at all times, and charge them off to

current expense. If so, the number of parts to include in this calculation declines considerably, making the measurement much easier to
calculate. The formula is:

$$\text{Percentage of new parts used in new products}$$
$$= \frac{\text{Number of new parts in bill of materials}}{\text{Total number of parts in bill of materials}}$$

A reasonable argument against this metric is that it discourages engineers from replacing more expensive parts, or parts that are not reliable. While valid, these changes can be documented and then removed
from the calculation.

ON-TIME PARTS DELIVERY PERCENTAGE

If suppliers can deliver parts on time, then a company can reduce its
safety stock to minimal levels to cover potential parts shortages. The
on-time parts delivery percentage must be calculated individually for
each supplier. To calculate it, subtract the requested arrival date from
the actual arrival date. If the intent is to develop a measurement that
covers multiple deliveries, then create an average by summarizing this
comparison for all the deliveries, and then dividing by the total number
of deliveries. Also, if an order arrives prior to the requested arrival
date, convert the resulting negative number to a zero for measurement
purposes; otherwise, it will offset any late deliveries, when there is no
benefit to the company of having an early delivery. Because a company
must pay for these early deliveries sooner than expected, they can even
be treated as positive variances by stripping away the minus sign.

PERCENTAGE OF RECEIPTS AUTHORIZED
BY PURCHASE ORDERS

The use of purchase orders is one of the best controls over unauthorized buying. To track the proportion of these purchases, track the percentage of authorized receipts. It may be useful to subdivide this
calculation by supplier, to see where problems are arising.

The receiving department should maintain a receiving log, on each line of which is recorded the receipt of a single product within an order. Using the line items in the receiving log that correspond to the dates within the measurement period, summarize the number of receipt line items authorized by open purchase orders by the total number of receipt line items in the log. The formula is as follows:

$$\text{Percentage of receipts authorized by purchase orders}$$
$$= \frac{\text{Receipt line items authorized by open purchase orders}}{\text{Total receipt line items}}$$

OBSOLETE INVENTORY PERCENTAGE

A company needs to know the proportion of its inventory that is obsolete, in order to set up a reserve and to determine the status of inventory dispositions. To calculate it, summarize the cost of all inventory items having no recent usage, and divide by the total inventory valuation. The amount used in the numerator is subject to some interpretation, since there may be occasional usage that will eventually use up the amount left in stock, even though it has not been used for some time. An alternative summarization method for the numerator is to only include those inventory items that do not appear on any bill of material for a currently produced item.

In summary, a few key metrics can give management a clear idea of the velocity of inventory, the accuracy of databases that affect inventory, obsolescence, new parts usage, and the cost to maintain inventory. This information can then be used to make decisions to improve inventory utilization, database accuracy, new product designs, or the cost-effectiveness of maintaining inventory.

SUMMARY

The main focus of this chapter has been that less inventory makes the accountant's job easier. A negligible on-hand inventory vastly reduces the risk of an incorrect inventory transaction. Also, accelerated inventory turnover eventually mandates the use of inventory "pull" systems that work better without most traditional inventory transactions—there is no receiving, no work-in-process tracking, and so on. Instead, inventory flows so swiftly that the only significant accounting transaction is to record what finished goods were produced. This streamlined state of affairs also yields less transaction errors and fewer controls. The key metrics to monitor the situation now become inventory turnover and bill of materials accuracy. However, few companies achieve such an enlightened state, because doing so requires the active cooperation of the engineering, purchasing, production scheduling, and warehouse departments for a prolonged period of time. Thus, the accountant is usually mired in an in-between state, tracking more inventory transactions and using more controls than are really necessary.

Chapter 5

Accounts Payable

The accounts payable function is heavily laden with a blizzard of transactions arriving in the accounting department from all directions: receiving transactions, purchase orders, and supplier invoices. The accounting staff must reconcile all of these incoming transactions, which frequently requires a substantial amount of research. In addition, despite the evidence of at least three different preexisting transactions regarding a purchase, the accounting department sends out the supplier invoice for approval! This sequence of transactions may take a whole month to process (no wonder most companies prefer 30-day payment terms—they can't pay any faster). In this chapter, we review the existing purchasing and payables system, and ways to reduce processing time while still maintaining control over the process.

CURRENT SYSTEM

A typical payables transaction involves many approval steps, takes an inordinate amount of time, and crosses department lines regularly. These factors result in perhaps the most confused paperwork flow of any transaction with which the controller must deal.

A purchasing request can begin in any department when an employee fills out a purchase requisition for an item, gets it signed by a manager, and brings it to the purchasing department. That department then adds more information to the requisition, such as the account number to be charged and the supplier to be used. This last item may require considerable time if the purchasing department puts all items over a certain price level out to bid.

The purchasing department then creates a purchase order and has it signed by a manager. These documents are numerically sequenced, and the purchase order stock is usually locked away when not in use. Copies of the purchase order go to the supplier as well as to the accounting and receiving departments, and another copy is filed in the purchasing department (usually with a copy of the requisition).

Once the ordered part has arrived, a receiving clerk inspects it and marks down the receipt in a receiving log. This log provides evidence that the item has been received, and it is needed by the accounting department as backup for paying the supplier; therefore, a copy of the log is sent to the accounting department.

The supplier then sends the company an invoice for the item just received. The payables clerk compares the quantity listed on the invoice to that on the receiving report and purchase order to ensure that payment is being made only for the item ordered and received. The prices, discounts, and terms of shipment are also reviewed to ensure that no overpayments are made. The purchase order, receiving document, and invoice frequently do not match and so must be reconciled. The payables clerk becomes an investigator, checking with the payables department for incorrect terms and prices, the receiving department for the amount received, the shipping company for evidence of shipment, and (most often) with management to see whose inbox currently contains the invoice that may require multiple layers of management approval. While this investigation goes on, the payables staff may accumulate a large number of unverified supplier invoices, which can build into a considerable backlog of work.

If suppliers send monthly statements, the payables staff reconciles them against in-house supplier records, and contacts the suppliers if there are any discrepancies. If suppliers do not send statements, then the only warning that a payables clerk will receive about a late payment is an angry call from the supplier.

After all documents for a purchase have been reconciled, the payables clerk authorizes payment to the supplier. A check is printed, signed, and mailed. All documents are stapled together, along with a copy of the check remittance, and filed away. A stripped-down version of this payables process, excluding management approvals and document filing, is shown in Exhibit 5.1.

One of the biggest problems with the typical accounts payable reconciliation process is that the payables clerk is often deluged with conflicting information from many sources. For example, the incidental information on the supplier invoice may not match the information on the purchase order—the tax rate or shipping and handling prices may vary. The purchase order quantity frequently varies from the amount received (and invoiced). The received amount is frequently miscounted or not recorded in the receiving log at all. If damaged goods are returned or there are overages, the supplier must be debited or credited for these variances. Finally, there may be a legal contract associated with the payment that must be consulted for such items as scheduled price increases and methods of delivery. In short, the payables clerk faces a monumental task that must be performed for even the smallest, most insignificant items; in fact, cheaper parts tend not to receive as much attention from the purchasing and receiving personnel as expensive items.

Another drain on accounting staff time is the processing of employee expense reports. As shown in Exhibit 5.2, an employee fills out an expense report, attaches receipts for larger items, obtains supervisory approval, and then sends it to the accounting department, which reviews it for compliance with company expense reimbursement policies. This process consumes a great deal of time, to the extent that an employee may not be reimbursed in time to pay for expenses within the payment time limit on his credit card.

Accounts payable also involves the disbursement of cash. A typical disbursement system generally requires the accounting staff to group invoices by payment due date, so that each supplier can be paid for multiple invoices with one check. Checks are then created, attached to

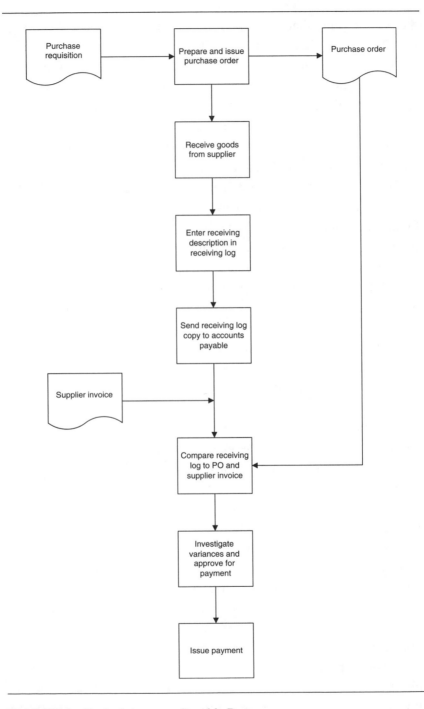

Exhibit 5.1 Typical Accounts Payable Process

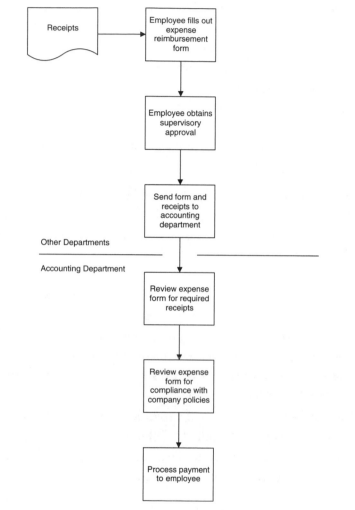

Exhibit 5.2 Employee Expense Reporting Process

the supporting documentation as evidence, and forwarded to a check signer. The check signer reviews the supporting documentation and (presumably) signs the check. If there is a problem, the check is returned unsigned, and the accounting staff investigates the problem. The mail room then mails the check to the supplier, while the accounting staff stamps or perforates the supporting documentation with a "PAID" stamp and files it.

Many businesses pay for incidental expenses with petty cash funds. The typical petty cash fund operates on an imprest fund basis, so the balance is fixed. At any time, the cash plus the unreimbursed vouchers should equal the amount of the fund. Numerous funds of this type may be necessary in corporate branch offices. A typical petty cash system generally follows these steps:

1. Obtain a signed voucher from the employee, written in ink (to prevent subsequent changes).
2. Issue cash equaling the amount of the voucher to the employee.
3. When the cash amount drops below a certain level, write a check made out to "petty cash" that returns the fund to its predetermined maximum level.

Finally, payables documents are very likely to become lost, because they are transferred across departmental boundaries so frequently. As shown in Exhibit 5.3, departmental transfers occur for all management approvals, purchase requisitions, purchase orders, and receiving logs. Every time such a transfer occurs, it is possible that a document will be delayed or lost.

The move and wait times in the traditional payables process greatly slow it down, and every time something is moved, the paperwork may be lost or misinterpreted. The value-added analysis shown in Exhibit 5.4 lists each step in the purchasing and payables process and

Exhibit 5.3   Payables Paperwork Movement Across Departments

	Accounting	Receiving	Purchasing	Legal	Management
Purchase requisition	√		√		√
Purchase order	√	√	√		√
Invoice	√				√
Receiving log	√	√			
Supplier debits/credits	√	√			
Contract information	√		√	√	

Exhibit 5.4 Purchasing and Payables Value-Added Analysis

Step	Activity	Time Required (Minutes)	Type of Activity
1	Employee prepares purchase requisition	5	Non-value-added
2	Employee moves to supervisor	1	Move
3	Supervisor signs requisition	1	Non-value-added
4	Employee moves back to office	1	Move
5	Employee files copy of requisition	1	Non-value-added
6	Employee moves to mailbox area	1	Move
7	Employee leaves requisition in purchasing mailbox	1	Non-value-added
8	Wait time for mail pickup	120	Wait
9	Purchasing person brings requisition back to purchasing department	1	Move
10	Purchasing person obtains order information	30	Value-added
11	Purchasing person completes a purchase order	10	Value-added
12	Authorized signer moves to authorized purchase order signer	1	Move
13	Authorized signer reviews and signs the purchase order	1	Non-value-added
14	Purchasing person moves back to office	1	Move
15	Purchasing person files copy of purchase order	1	Non-value-added
16	Purchasing person moves to mailbox area	1	Move
17	Purchasing person mails copy of the purchase order	1	Non-value-added
18	Purchasing person leaves copy of purchase order in requester's mailbox	1	Non-value-added
[Parts are received from supplier]			
19	Receiving person enters receipt in receiving log	5	Non-value-added
20	Receiving person moves to the mailbox area	1	Move

(Continued)

Exhibit 5.4 Continued

Step	Activity	Time Required (Minutes)	Type of Activity
21	Receiving person leaves bill of lading in mailbox of accounting department	1	Non-value-added
22	Wait time before accounting person retrieves mail	120	Wait
23	Accounting person brings bill of lading back to the accounting department	1	Move
24	Accounting person files bill of lading	1	Non-value-added
[Supplier invoice arrives]			
25	Mail room person leaves invoice in accounting department mailbox	1	Non-value-added
26	Wait time before accounting person retrieves mail	120	Wait
27	Accounting person brings invoice back to the accounting department	1	Move
28	Accounting person logs invoice into computer system	5	Non-value-added
29	Accounting person takes invoice back to mailbox area	1	Move
30	Accounting person leaves invoice in mailbox of authorized approver	1	Non-value-added
31	Wait time until authorized approver retrieves mail	120	Wait
32	Authorized approver signs invoice and places in mailbox of accounting department	2	Non-value-added
33	Wait time until accounting person retrieves invoice	120	Wait
34	Payables clerk retrieves copy of bill of lading	1	Non-value-added
35	Payables clerk retrieves copy of purchase order	1	Non-value-added
36	Payables clerk matches purchase order to bill of lading and invoice	5	Non-value-added

37	Payables clerk files in one bundle the purchase order and bill of lading	1	Non-value-added
38	Payables clerk moves to mailbox area	1	Move
39	Payables clerk leaves invoice in mailbox of authorized signer	1	Non-value-added
40	Wait time until authorized signer retrieves invoice	120	Wait
41	Authorized signer reviews and signs invoice	2	Non-value-added
42	Authorized signer leaves approved invoice in accounting department mailbox	1	Non-value-added
43	Wait time until accounting person retrieves mail	120	Wait
44	Accounting person brings mail back to accounting department	1	Move
45	Payables clerk retrieves the purchase order and bill of lading and matches with invoice	5	Non-value-added
46	Payables clerk enters approval code into computer system for payment by check	2	Non-value-added
47	Accounting clerk prints checks	60	Value-added
48	Payables clerk matches checks to backup packet of purchase order, invoice, and bill of lading	60	Non-value-added
49	Accounting clerk takes checks and packets to authorized check signer	1	Move
50	Authorized signer reviews packets and signs checks	10	Non-value-added
51	Accounting clerk takes checks back to accounting department	1	Move
52	Payables clerk removes backup packets from checks, attaches copy of check to packets, and files packets	2	Non-value-added
53	Accounting clerk brings checks to mail room	1	Move
54	Mail room staff mails the checks	2	Value-added

the time required to complete it. A value-added item is considered to be one that brings the purchasing or payables transaction closer to a conclusion.

Exhibit 5.5 shows that only four of the steps bring the purchasing and payables transactions closer to a conclusion; the remaining steps

Exhibit 5.5 Summary of Purchasing and Payables Value-Added Analysis

Type of Activity	No. of Activities	Percentage Distribution	No. of Hours	Percentage Distribution
Value-added	4	7%	1.70	9%
Wait	7	13	14.00	78
Move	16	30	0.30	2
Non-value-added	27	50	2.02	11
Total	54	100%	18.02	100%

are related to moving paperwork from person to person, getting approvals, or making copies for filing purposes. In terms of time required, the value-added steps can be concluded in 1.7 hours, while the moving, waiting, and non-value-added portions of the transaction take up about two business days. Thus, the actions needed to conclude the transaction are only a small part of the total process.

In summary, a typical accounts payable system requires considerable reconciliation work by the accounts payable staff to determine the amount to pay a supplier, based on sometimes conflicting information from a large number of separately maintained databases. This results in a long processing time and a high clerical cost per transaction. The following section reviews several ways to reduce the transaction volume and the amount of reconciliation work, resulting in faster completion of the payables process.

REVISED SYSTEM

The typical purchasing and payables system is burdened by an excessive number of approvals, too many forms requiring reconciliation, and a high volume of items requiring processing. This section examines methods for improving and speeding up the purchasing and payables process.

An important step is *reducing the volume of transactions handled.* An analysis of the dollar value of each payables transaction will reveal that a disproportionate amount of labor is devoted to very small purchases that are not worth all the approvals, purchase requisitions, purchase orders, supplier invoices, matching, and payment with individual checks. Instead, institute the use of corporate purchasing cards. These credit cards allow employees to purchase small-dollar items directly from suppliers. The advantage to the accounting department is that a large number of transactions are reduced to a small number of monthly credit card payments. A typical purchasing and payables transaction costs about $30 to process; by switching to corporate credit cards, the cost per transaction drops to less than $1. Also, by reducing the overall number of payables transactions, the payables staff can concentrate on the remaining transactions, which usually results in fewer transaction mistakes. Furthermore, having corporate credit cards makes it less necessary to maintain a petty cash fund, which in many cases can be eliminated.

Another way to reduce the volume of transactions handled by the payables staff is *simplifying the ordering process for commonly purchased items*. For example, most office supplies are ordered through one office supply company; that being the case, it is easy to create an order form for the most commonly ordered materials and post it in the company's supply room. A periodic review of the supply room will reveal which items are running low; those items can be noted on the order form and faxed to the office supply company for rapid refilling. Many office supply companies even deliver the desired materials free of charge, so no transportation arrangements are needed. The same process can be used for manufacturing supplies, although they are not usually ordered from just one supplier. In either case, the ordering process avoids the usual purchasing and payables process involving purchase requisitions, purchase orders, and multiple management approvals. If a corporate purchasing card is used to automatically pay for the supplies, the controller can also avoid matching receivables documentation to supplier invoices. Since supplies involve moderate dollar amounts but very large transaction volumes, the risk of losing large amounts of money is minimal and the volume of payables transactions is noticeably reduced.

Another way to reduce transaction volume is *automating the expense report processing function*. This system allows employees to enter their expense report information directly into a central computer system, which collects information about specific days being reported, expense items requiring receipts, and explanations for missing receipts. Then the system prints out a transmittal slip, to which the employee attaches all receipts needed for the expense report. The employee gives the transmittal slip and attached receipts to the accounting department, which only has to verify that all the receipts listed on the transmittal slip are attached. The electronic expense report is sent over the computer network to the employee's supervisor for approval. After the electronic approval and the receipts have been received, the accounting department issues an Automated Clearing House (ACH) electronic payment to the employee. A flowchart of the process is shown in Exhibit 5.6. This solution requires an expensive and lengthy programming effort. For those requiring a faster solution, several companies offer the service over the Internet.

An alternative way to reduce the review work associated with expense reports is *reviewing expense reports only by random audit*. This greatly reduces the detailed examination currently used in nearly all companies. Though it is likely that employees will be able to charge off some expenses when an expense report is not selected for an audit, the prospect of being audited will act as a deterrent for most employees.

Many invoices are for negligible amounts, and so present little risk to a company if they are paid without any supervisory approval at all. The controller can *set a dollar limit below which all invoices are automatically approved* for payment. Anything smaller is simply not worth the review cost. In case many small invoices are being used to circumvent the approval threshold, a report showing the total number of payments made to suppliers will indicate such situations.

Another way to avoid approvals is called *negative assurance*. Under this technique, invoice copies are sent to authorizing employees, and are automatically paid when due unless the employees tell the accounts payable staff *not* to issue payment. By focusing only on

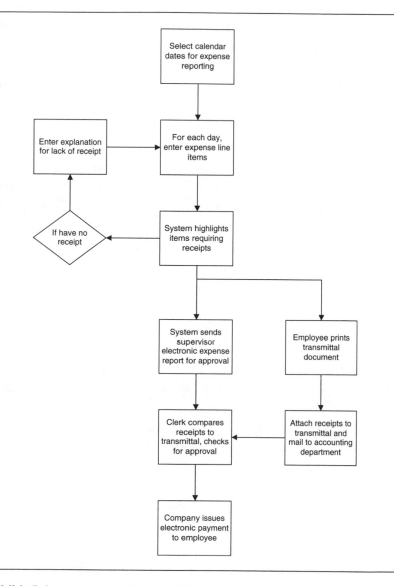

Exhibit 5.6 Automated Employee Expense Reporting System

those invoices that may be incorrect, the accounting staff can process the vast majority of all submitted invoices without cajoling anyone to submit an approved document.

It is possible to reduce the amount of paperwork associated with each remaining transaction. The first step is *paying from the purchase order.* This has multiple benefits of one less piece of paper to handle (the supplier invoice), one less required management approval, and one less item to include in the document-matching process. Under this system, all open purchase orders are available online at the receiving dock. When goods arrive, the receiving clerk accesses this database and checks off the quantity received against the amount authorized. The computer should incorporate a pre-set tolerance level for closing purchase orders even if the amount received is off slightly. The system then schedules a supplier payment. In short, payment authorization has moved to the receiving dock from the accounts payable department.

It is also possible to shorten the length of the overall payables cycle by *eliminating purchase requisition approvals.* The bulk of all purchase orders written are for items needed to manufacture products; these purchases are frequently issued automatically by a material requirements planning (MRP) system or are issued manually based on pre-approved bills of material. These items require no approval, since the information on which the purchase is based (the bill of materials) has already been approved by management. The next-largest item normally requisitioned is supplies (either for the office or the manufacturing area). With the use of corporate purchasing cards, these transactions no longer require requisitions.

The most advanced level of efficiency is to pay suppliers based on production records instead of supplier invoices or (as noted earlier) purchase orders. This system is called *evaluated receipts.* The key item here is to avoid the review by receiving personnel of the incoming shipments from suppliers. Instead, the parts are moved directly to the manufacturing area, where they are used immediately. Then, the amount of each part (as listed on the bill of materials) is multiplied by the number of products completed to arrive at the total number of parts the company has received from the supplier. This unit total is then multiplied by the cost per unit, as noted in the purchase order, thus arriving at a total amount to pay the supplier. Of course, the system must also include reporting for items damaged during production, so that the supplier is paid for these items as well. The use of payments based on production records is only possible in a well-run just-in-time manufacturing system, because the payment scheme will not work unless the following factors are present:

- Only enough parts are delivered to manufacture the product. If excess parts are delivered, additional accounting is required to determine the number of units for which to pay the supplier; in this case, it would be easier from purchase order information than from production records.

- Production is rapid. If it takes an inordinate amount of time to manufacture a product, the supplier must wait too long for payment, since payments are based on finished products.

- Only one supplier supplies each part. It is very difficult to pay suppliers when more than one supplies the same part. When payments are determined based on the parts included in a finished product, there is no way to tell which supplier delivered an included part.

- The bills of material are totally accurate. If they are not, suppliers will not be paid for the correct quantity of parts delivered.

- Product changes are minimal. If the quantity of a part used in the product is constantly changed because of design iterations, it is very difficult to determine the correct number of parts for which to pay the supplier.

- Bulk purchases are not needed. If bulk purchases are required for key items that can only be procured in volume, because of pricing, packaging, distances traveled, and so forth, then they must be paid for by some other means than production records, since it may be some time before the company would otherwise issue payment.

An evaluated receipts system is shown in Exhibit 5.7.

An issue for larger companies to consider is *reducing the number of accounts payable systems to one.* Many large companies have multiple systems, usually because acquired companies were allowed to retain their existing systems.

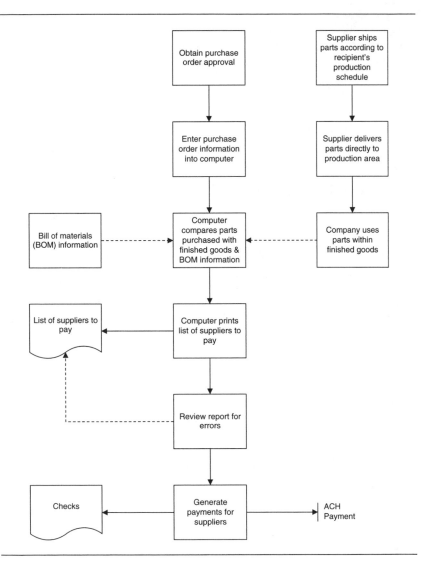

Exhibit 5.7   Evaluated Receipts System

Even if a company chooses to retain its existing accounts payable system, it can still *improve payables information accuracy*, so there are fewer errors to reconcile. One method is to negotiate with suppliers to prominently display the delivery quantity on inbound shipments, or to package deliveries in easily countable containers, such as units of ten. Also, to keep items from mistakenly not being recorded in the receiving log, set up an online system at the receiving dock, where employees must reduce open purchase order amounts at the point of receipt.

Suppliers themselves can be used to correct another problem, which is entering a supplier invoice into the computer system under an incorrect supplier identification code. To avoid this problem, the accounting staff should send out a mailing to all suppliers, informing them that the address of the accounts payable department has changed. The new address should include a *"to the attention of" code for each supplier.* This will result in supplier invoices arriving that already contain the correct identification number, and which will virtually eliminate the miscoding of invoices to the wrong account.

The last two recommendations involved working with suppliers to make improvements. The best possible example of supplier training is to issue to the entire supplier base a *supplier manual.* The manual contains a great deal of information that assists in the development of smoothly interlocking billing and payables systems between each supplier and the company. The manual can describe where invoices are to be sent, where a purchase order reference should be noted on their invoices, when the company issues payments, access codes for any online inquiry system, special instructions for an evaluated receipts system, and so forth.

In summary, a variety of techniques are available for increasing the speed of accounts payable transactions. To accomplish this transformation safely requires considerable attention to controls, which are addressed in the next section.

CONTROL ISSUES

Accounts payable requires many controls, since missing ones can result in major expenditures that may not be authorized by management. For example, if the controller were to eliminate the comparison of purchase orders to invoices by the accounts payable staff, then suppliers could charge excessive rates without anyone noticing. Similarly, eliminating the comparison of supplier invoices to the receiving log would allow suppliers to bill for parts never sent. This section contains examples of what would happen if various payables controls were removed in order to increase the speed of accounts payable transactions, and suggests possible solutions to the resulting control problems. These examples pertain to the changes suggested previously for a revised payables system.

- *Stop requiring approval of supplier invoices.* If supplier invoices are approved by management, this is a last-chance review of payments about to be made. If not used, the next most recent chance to catch an improper payment is the review of purchase orders, while manually signing checks can also catch errors.

  A basic controls rule is that, if one control is removed, strengthen the remaining ones. The reason why so many approvals are required throughout the payables process is that many accounting systems are so leaky that payments slip through individual layers of approval, and so can only be caught by setting up a web of approvals. The solution is to tighten control over the key approval point—the purchase order. If everything of material size requires one, and a manager must approve it, then there is no further need for downstream control.

  Enforcing this control point involves rejecting items received that have no purchase order, educating all employees about this requirement, and conducting frequent audits of purchase orders to ensure that the system works. It may also be necessary to pre-number purchase order forms and store them in a secure location.

  Service agreements for such items as janitorial services and telephone repairs are frequently negotiated outside of the purchasing department. Also, services companies do not enter the company through the receiving dock, so there is no documented way to determine if they've performed work. Consequently, a purchase order control does not work well for services; instead, the primary control retains management approval of the supplier invoice.

  Some companies negotiate long-term contracts for key supplies. For example, a power plant may have a 20-year contract for coal.

In such cases, create a blanket purchase order that is reviewed periodically to ensure that any changes required by the contract are enforced in the payment system.

- *Stop requiring approval of purchase requisitions.* If management must sign off on purchase requisitions, managers have an early control over purchases, which is later reinforced by management review of purchase orders before they are issued to suppliers. It also reduces the workload of the purchasing staff, who may otherwise source a part and prepare a purchase order only to find that management disallows the purchase during its review of outgoing purchase orders. If this control were removed, then a leaky purchase order review might let inappropriate purchase orders be issued.

    The control can be eliminated as long as there is an extremely strong downstream control that requires approved purchase orders for all significant purchases.

    Another option is to purchase minor items from a select list of pre-approved suppliers, and have employees use a standardized form for placing orders for specific items. Since not much money is involved, there is no need for management approval. A sample order form is shown in Exhibit 5.8.

    If a company starts using corporate purchasing cards, the number of purchase requisitions will be reduced, by allowing employees to purchase small-dollar items directly from selected suppliers. Of course, the potential exists for loss of control over purchases, since a credit card can be used to acquire inordinate amounts of supplies. This issue can be offset by using cards with maximum expenditure limits, and frequent reviews of purchases made.

- *Stop paying based on supplier invoices.* If a company does not pay from supplier invoices, differences between the payments made by the company and payments expected by the supplier will occur with greater frequency, requiring substantial time by the accounting staff to reconcile.

    The contract between the company and the supplier is the purchase order, not the invoice. As the base-level document, the

Exhibit 5.8   Office Supplies Order Form

OFFICE SUPPLIES ORDER FORM

Date _____

Name of Person Ordering _____

Description	Part Description	Supplier Part No.	Quantity
Fasteners			_____
¼″ Alligator clips	ClipArt ¼″	CA-2500	_____
½″ Alligator clips	ClipArt ½″	CA-5000	_____
Paper			_____
White legal	Mead White Long	ME-4320	_____
Yellow legal	Mead Yellow Long	ME-4321	_____
Staplers			_____
Stapler, regular	SwingLine Regular	SW-100A	_____
Stapler, heavy duty	SwingLine Heavy	SW-100B	_____
Tape			_____
¼″ tape	Scotch ¼″ tape	3M-1352	_____
½″ tape	Scotch ½″ tape	3M-1353	_____

purchase order with the price noted on it is legally binding. Thus, the purchase order should be used to pay the supplier. The company can dispense with invoices and match the receiving log to the purchase order, paying the supplier the total of the quantity received times the price on the purchase order. Alternatively, the company can pay the supplier based on the quantity used in the production process (based on the production schedule and the bill of materials) and dispense with the receiving log entirely. Though this system avoids considerable accounting labor, several problems may arise with respect to pricing or quantity disputes, supplier debits and credits, shipping charges, and taxes.

The supplier may sometimes charge a price that varies from the amount on the purchase order. The company's position should be that it will pay according to the price listed on the purchase order; and changes to that price must come from a purchase order change approved in advance by management. Also, the supplier may complain that the company is not paying for the quantity that was shipped. This problem arises when either the supplier or the company miscounted the items shipped. One solution is to rely on the documents of the shipping company as the final arbiter. If so, the company and supplier should jointly devise packaging that is easily counted.

Shipping fees are charged for nearly all deliveries. Since the company is paying from a purchase order, it has no way of knowing what the shipping charge is for each delivery. One way to avoid this problem is to have the shipper charge the company rather than the supplier, thereby avoiding any markups that a supplier might add onto the shipping fee. Another method is to negotiate a standard shipment fee with the supplier in a long-term contract. Then the company can factor the standard charge into its purchase orders and pay the fee based on the purchase order information.

Finally, taxes must often be paid on parts, and the supplier usually provides the tax information on the supplier invoice. If the invoice is being ignored, the company must get the tax information from other sources. In many commercially available accounting packages, tax tables are available by state, county, and city, which allows a company to make tax payments without supplier-provided information.

- *Stop three-way matching.* Purchase orders are matched to invoices and the receiving log in order to verify that the items ordered were received, and that the correct price is paid. If this matching control were removed, the company might make payments even if materials were not received and might use an incorrect unit price.

If the supplier delivers parts directly to the manufacturing area in the exact quantities needed for production, the production schedule can be used as evidence of materials receipt. The quantity of goods produced are multiplied by the number of parts required for each unit; this part quantity information is listed on the bill of materials. Then the price listed on the purchase order is multiplied by the total number of units used to derive the amount to be paid to the supplier. Of course, any parts scrapped during the manufacturing process must be carefully recorded and credited back to the supplier.

- *Stop signing checks.* This is the last control before a check is mailed to the payee, since the check signer is assumed to be reviewing the checks and any accompanying documentation. If checks are not manually signed (e.g., being printed through an imprinting device instead), it is possible for checks to be improperly issued to the wrong parties. The best alternative to this control is stronger enforcement of purchase order approvals, so that only very small-dollar invoices are paid without some form of prior approval.

Some controls should be continued, because they are cost/effective in all environments, and because there is a significant likelihood of increased losses if they are eliminated. Several highly recommended ones are noted below:

- *Account for the sequence of purchasing documents.* When the purchase order is the primary means by which both quantities and costs are determined, it is crucial that the form itself be controlled, so that forms cannot be removed and used to illegally purchase materials in the name of the company. To do so, lock up all unused purchase orders, and maintain a log of those used. Review the log regularly to ensure that all forms are accounted for.

- *Audit expense reports.* Review a small selection of employee expense reports to ensure that only appropriate charges are being made to the company. A few audits will reveal that some employees always report expenses fairly, while others always stretch the company rules to report the maximum possible amount. Once patterns emerge, the selection method should concentrate on those employees whose reports continue to reveal reporting problems, and review other employees' reports less often.

- *Mutilate paid invoices.* It is relatively common for invoices to be paid more than once. To avoid this problem, invoices should be marked or mutilated in some manner that clearly identifies them as having been paid.

- *Link travel policies to expense reports.* For those companies with the funding to invest in automated expense reporting systems, an additional system feature to consider is the storage of corporate travel policies in a rules database that can be cross-referenced by the expense reporting system. By doing so, any expenditures that break company policy, such as first class air travel, movies, or clothing purchases, will be automatically flagged and rejected.

- *Review vouchers for duplicate submissions.* Employees can re-submit vouchers for reimbursement in successive expense reports. This can be controlled by reviewing voucher dates for old vouchers that could have been resubmitted with previous expense reports and then checking previous reports for attached copies of those vouchers.

- *Fill out petty cash vouchers in ink.* An employee can alter the amount on a petty cash voucher and pocket the difference. Requiring that vouchers be filled out in ink makes the forging task more difficult. An auditor can then periodically review the vouchers for alterations.

- *Report receiving exceptions to accounting.* Any changes in the amount received from the amount ordered should be carefully noted at the receiving dock (if receiving is still used) and communicated to accounting. The best method is for the receiving staff to call up related purchase orders on a computer terminal as goods arrive, and punch in variances from the ordered quantity on the spot.

- *Send freight documents to accounting.* Even if freight is paid by the supplier, a company should still retain freight documents, because they provide evidence of receipt. If the company has a dispute with a supplier over whether a shipment was received, the shipping documents will be available as evidence.

- *Review long-term contracts.* There is a potential for incorrect payments when long-term contracts are used, because fees may change over time and may not be noted by the accounts payable department. It is useful to create a summary of each contract, which is part of a monthly review of all contracts.

- *Complete a daily bank reconciliation.* A frequently-updated bank reconciliation based on a bank's online banking information is an excellent way to spot payment irregularities, such as fraudulent checks. This is an after-the-fact control, since losses will have already occurred, but it also spotlights areas in which new controls are needed.

- *Review credit card documents.* Although corporate purchasing cards are not intended for large-dollar purchases, the total volume of low-dollar purchases for which they may be legitimately used is so large that the overall dollar volume involved can be considerable. For that reason, someone should review the credit card statements to determine the types of items purchased, the purchase locations, and the dollar amounts involved. It may be possible to download credit card information from the card provider, making it easier to sort through purchasing information.

- *Fill in empty spaces on checks.* There is a danger that checks can be altered after they are signed, so that the amounts paid are more than they should be. This can be avoided by having all checks printed by a computer, which automatically fills in all empty slots on the payment line with a meaningless symbol, such as an asterisk.

- *Lock up the signature plate.* If a signature plate is used to automatically affix a signature to checks, then it can be used to create improperly signed checks. Therefore, always lock it up when it is not in use.

- *Use positive pay.* Under this approach, a company creates a file containing the check number, date, and amount of all checks it issues each day, and forwards the file to its bank. The bank then compares the check information in this file to checks being presented for payment, and refuses to accept any checks containing different information.

- *Use debit blocking.* It is possible for a third party to use an ACH debit transaction to illicitly remove money from a company's bank account. To avoid this, require the bank to install debit blocking on the account. This can be fine-tuned, so that only the debits of certain suppliers are allowed, and further within a specified date range and dollar amount.

- *Mutilate voided checks.* There is a risk that any voided check will be improperly cashed. A VOID perforation or a similar mark will eliminate this risk.

- *Use periodic audits.* The internal audit department can potentially review a number of areas within the accounts payable function. Here are some possibilities:

  - *Review duplicate payments.* This can be checked by periodically tracing payments back to their supporting receiving or freight documents.

  - *Review supplier returns.* A company may have trouble recording supplier credits for parts returned to suppliers. This can be discovered by tracing shipping documents back to the accounting department to ensure that proper credit has been recorded.

  - *Compare authorized to paid cost.* Without proper verification, a company can pay more from a supplier's invoice than was specified in the purchase order. Auditors can compare the two documents to find exceptions.

  - *Compare amount received to amount paid.* A company may pay for the quantity listed on the supplier invoice, rather than the amount received. Auditors can compare the two documents to find exceptions.

  - *Compare receipts to payable records.* A potentially major issue is under-recorded accounts payable. To discover such items, auditors can compare the quantities received (and their related costs as listed on purchase orders) to the costs recorded in the accounts payable records.

  - *Review bill of materials accuracy.* For those companies paying suppliers based on the number of parts used in its finished products, it is critical that their bills of material be totally accurate—otherwise, payments to suppliers will be incorrect. Auditors can compare bill of material records to actual usage and note any differences.

o *Review purchase order numbering sequence.* When paper-based purchase orders are used as the primary approval for payables, control over the forms are critical, since illicit orders can be placed with stolen forms. Thus, auditors can review the sequencing of purchase orders to see if any are missing.

o *Review service charges.* One of the most difficult areas to verify is payment for services, since there is no supporting receiving documentation. One possibility is to compare payments to sign-ins by service personnel in the company visitor log, and to verify the approval of supplier invoices by company managers.

o *Review petty cash.* Employees can take money from petty cash with the intention of returning it later. This can be spotted by auditing petty cash without notice, so that missing money cannot be returned to the cash fund before the audit has been completed.

In summary, the very slow accounts payable process can be speeded up more than any other accounting function. This can be accomplished by eliminating a number of approvals, paying from purchase orders, using purchasing cards, and automated employee expense reporting systems. While these new systems can dispense with a number of traditional controls, the remaining controls should be robust, and supplemented by a number of periodic internal audit reviews.

COST/BENEFIT ANALYSIS

This section describes how to write cost/benefit analyses for corporate purchasing cards, automated expense reimbursements, paying from purchase orders, skipping purchase requisition approvals, using evaluated receipts, and reducing the number of payment approvals. In these examples, expected costs and benefits are as realistic as possible.

The justifications are reduced to a one-year time frame in order to reduce the size of the examples; in reality, net present value calculations for five years of costs and benefits are more realistic.

USE CORPORATE PURCHASING CARDS

You want to introduce corporate purchasing cards at your company, and need to present a cost/benefit analysis to the purchasing manager. You find that the annual fee of each of 20 corporate purchasing cards is $20. You also conduct an analysis of all company purchases and find that 20% of all purchase orders are for purchases of less than $100, with an average purchase price of $52. Of those purchases, 75% are from five local supply companies, all of which accept credit cards as payment for goods. The company issued 28,000 purchase orders in the previous year, totaling $33,600,000. An activity-based costing analysis has shown that the average purchase order, including all related activities, requires $30 to complete. A local internal auditors trade association has conducted a confidential survey of employee theft using corporate purchasing cards, and concluded that improper use increased the cost of items purchased with corporate purchasing cards by 0.5%. Is it worthwhile to use corporate purchasing cards? The analysis follows:

Cost of Using Corporate Purchasing Cards

Number of corporate purchasing cards	20
Annual fee per card	× $20
Total annual fee for cards	$400
No. of purchase orders	28,000
Percentage under $100	× 20%
No. of purchase orders under $100	5,600
Percentage of under-$100 POs allowing credit	× 75%
No. of under-$100 POs allowing credit	4,200

(Continued)

Average amount of under-$100 POs	× $52
Amount of POs converted to credit cards	$218,400
Expected percentage of improper credit purchases	× .5%
Total amount of improper credit purchases	$1,092
Total cost of using corporate purchase cards	$1,492
Cost of Not Using Corporate Purchasing Cards	
No. of purchase orders	28,000
Percentage under $100	× 20%
No. of purchase orders under $100	5,600
Cost to process a purchase order	× $30
Total cost to process under-$100 POs	$168,000

The analysis reveals that the costs associated with using corporate purchasing cards are minuscule compared to the savings resulting from their use. Based on the example, the conversion would pay for itself in less than three days.

INSTALL AUTOMATED EXPENSE REIMBURSEMENT SYSTEM

Giles von Strohe, CFO of the Near Beer Company, wants to install an automated employee expense reimbursement system. He asks you to explore the costs and savings of such a system and to recommend a course of action. You find that the company's 65 salespeople submit expense reports every week for an average of 40 weeks per year, which is a total of 2,600 expense reports per year. An accounts payable clerk reviews each expense report to see if expenses claimed are in accordance with company policies, matches receipts to the expenses listed on the summary sheet, and resolves discrepancies by calling the salesperson. The typical expense report review requires 15 minutes. The average payables clerk wage, fully loaded, is $17.50. Also, the sales manager reviews each expense report for six minutes, which may include a phone conversation with the submitting salesperson. The sales manager's salary is $120,000. Should the system be installed? The analysis follows:

Cost of Manual Expense Reimbursement System	
No. of expense reports	2,600
Clerical time to review report	× .25 hr
Total clerical hours	650 hrs
Cost/hour for clerk	$17.50
Total cost of clerical review	$11,375
No. of expense reports	2,600
Sales manager time to review report	× .10 hr
Total manager hours	260 hrs
Cost/hour of manager	$57.69
Total cost of manager review	$15,000
Total cost of expense report review	$26,375

Further investigation reveals that the company can purchase a bolt-on module for its existing accounting system that provides automated expense reporting functionality for $30,000. It must also create a user guide and train the sales staff in how to use the new application. Each salesperson will require one hour of training. The average salesperson earns $85,000 ($40.87 per hour). Additional analysis follows:

Cost of Automated Expense Reimbursement System	
Software cost	$30,000
Average cost/hour/salesperson	$40.87
Training time	× 65 hrs
Total training cost	$2,657
Total implementation cost	$32,657

In short, your investigation reveals that an automated system will save $26,375 per year on an ongoing basis, but will require $32,657 of up-front costs to implement. The payback is 1.2 years. An additional consideration if the analysis were to encompass future years would be the cost of software maintenance for the new software module.

PAY SUPPLIERS FROM PURCHASE ORDERS
RATHER THAN INVOICES

In an effort to cut costs, the CFO of Daisy Baby Foods wants to start paying suppliers from purchase orders rather than from supplier invoices. As the controller, you are assigned the task of creating a cost/benefit analysis. You find that the requirements definition, programming, testing, and documentation work related to the changeover will require the time of three programmers for half a year. The average programmer is paid $68,000. In addition, the mailing and staff costs associated with notifying key suppliers of the changeover will be $25,000. Training time for the staff that will be using the new system will cost about $17,000. You also estimate that two accounts payable clerks will be needed even after the new system is installed, because there will still be a limited number of supplier invoices arriving that do not have related purchase orders; in addition, the clerks must resolve payment disputes with suppliers, which will occur no matter what payment method is used. Despite these issues, the new system should allow the company to reduce the accounts payable department by four positions; the average salary of those positions is $35,000. Is this a cost-effective project? The analysis follows:

Cost to Implement a Payment System Based on Purchase Orders	
Cost/year per programmer	$68,000
No. of programmers required	× 3
Six months' work	× .50
Total programming cost	$102,000
Mailing and notification cost	+$25,000
Training cost	+$17,000
Total implementation cost	$144,000
Cost Reductions with New System	
Cost/year of payables clerk	$35,000
Clerical positions eliminated	× 4
Total salary-related savings	$140,000

The one-time cost to install the purchase order-based payment system is $144,000, versus annual savings of $140,000, which is a payback of just over one year.

ELIMINATE PURCHASE REQUISITION APPROVALS

A re-engineering team at A.L. Skrupp Corp. has recommended that purchase requisition approvals be eliminated. The controller, Mr. Boldrup, reviews the situation and finds that the average company manager can review, sign, and forward a purchase requisition in two minutes. The average time required to find a manager who can sign the requisition is one day. Discussions with a sampling of company managers reveals that they typically reject 2% of all purchase requisitions sent to them for approval. They are later called upon to sign off on completed purchase orders that contain the same information. The somewhat annoyed purchasing manager informs Mr. Boldrup that it takes 45 minutes of purchasing department time whenever it converts a purchase requisition into a purchase order; if an unapproved requisition is later rejected in purchase order form by a manager, then that is a waste of 45 minutes of purchasing time. The average company buyer earns $22.00 per hour. The average company manager earns $40 per hour. The purchasing department receives 82,500 purchase requisitions per year. Should the A.L. Skrupp Corporation eliminate manager sign-offs on purchase requisitions? The analysis follows:

Cost of Retaining Purchase Requisition Approvals	
Cost/hour of manager	$40
Minutes/hour	/60
Cost/minute of manager	$0.67
Time to review requisition	× 2 min.
Cost/requisitions of management review	$1.34
No. of requisitions per year	82,500
Total cost of management review	$110,550

(Continued)

Cost of Eliminating Purchase Requisition Approvals

No. of requisitions approved/year	82,500
Percentage of requisitions not approved	× 2%
No. of requisitions not approved	1,650
Time to create purchase order	× .75 hr
Total time to create purchase orders	1,238 hrs
Cost/hour of buyer	$22.00
Total cost of creating unneeded POs	$27,236

The analysis reveals that reviewing all purchase requisitions is costing the company a startling amount of management time, which is four times more expensive than the added cost to the purchasing department caused by extra requisitions being converted into purchase orders. An objection to this analysis is that the company does not actually realize any cost savings, since the managers will still be paid by the company, while the purchasing department may have to add a buyer to handle the extra purchase requisitions. However, this practice keeps some clerical work away from the management group, which will therefore have more time for value-added activities.

USE EVALUATED RECEIPTS

The CFO of the Shine Bright Lamp Company is interested in paying suppliers from the information contained in the company's production records, rather than from supplier invoices. The CFOs assistant, Mr. Dunwoody, is asked to construct a cost/benefit analysis of the project. He notes that one-half of all payments based on supplier invoices could be switched to payments based on production records. This would allow the company to cut its accounts payable department in half. The department has a staff of six, who are paid an average of $35,000 each. In addition, Mr. Dunwoody finds that the programming cost of paying from production records is substantial—four programmers will be required to design, create, and test the needed software over a period of six months. The company's average rate of pay for programmers is $82,000.

Also, the 18 key suppliers who will be affected by this change must be visited and informed of the new payment method. The company will send two programmers to the supplier locations to discuss the changes in their systems that will be required. Those two people must visit supplier's full-time for six months. Also, the purchasing manager will visit all 18 companies in advance during a one-month road trip to prepare the suppliers for the change. The purchasing manager earns $90,000 per year. Total travel costs for the programmers and purchasing manager will be $60,000. Finally, the company's bills of material (BOMs) are not yet accurate enough, so an engineer must be hired who will review the BOMs on a continuing basis. The salary of the engineer is expected to be $70,000. Should this project be implemented? The analysis follows:

Cost of Payment from Production Records	
Cost/year of programmer	$82,000
No. of programmers	× 6
Six months' work	× .50
Total programming cost	$246,000
Cost/year of purchasing manager	$90,000
One months' work	× 8.3%
Total purchasing manager cost	$7,470
Total travel cost	+ 60,000
BOM engineer salary	+ 70,000
Total implementation cost	$383,470
Benefit of Payment from Production Records	
Cost/year of payables clerk	$35,000
Clerical positions eliminated	× 3
Total savings	$105,000

It is evident from the preceding information that the payback period for this project would be too long for many companies. The economics of the case study would have been better if a larger number of payables clerks were made redundant (which is why this system is usually only implemented by large companies), or if a just-in-time manufacturing system were already in place. With a JIT system, BOMs are already extremely accurate, and there are fewer suppliers. For example, if there were half the number of suppliers and no need for an extra BOM engineer, the project cost would have dropped by about $100,000. In short, efficient manufacturing systems make it much easier to justify an evaluated receipts system.

REDUCE THE NUMBER OF PAYMENT APPROVALS

A key supplier has called the president of Tender Tissues, a maker of non-allergenic tissue paper, and complained that payments are extremely late. The president informs Mr. Underling, the controller, that this is unacceptable. Investigation by Mr. Underling reveals that several suppliers have been paid so late that they have forced Tender Tissues to accept C.O.D. terms, which carry a 5% surcharge; these payments account for 10% of the company's total annual purchases of $2,600,000. Internally, payments are made as soon as approved invoices are returned by supervisors to the accounts payable department. However, the time required to obtain signatures on invoices can take up to a month, since supervisors travel constantly. The solution appears to be the elimination of supervisor approval of supplier invoices and the strict enforcement of purchase order approvals. To enforce purchase order sign-offs, Mr. Underling estimates that a staff person (annual salary $38,000) must be assigned to the task full-time for four months. In addition, an internal auditor (annual salary $58,000) must review the purchase order controls for two weeks every year to report on any deviations from the control system. Is eliminating control over invoices worthwhile to the company? The analysis follows:

Cost of Enforcing Purchase Orders	
Cost/year of staff	$38,000
Fourth months' work	/3
Cost of staff	$12,667

Cost/year of auditor	$58,000
Two weeks' work	× .038
Cost of auditor	$2,204
Total salary cost	$14,871
Benefit of Stopping C.O.D. Payments	
Purchases/year	$2,600,000
C.O.D. payments	× 10%
C.O.D. payments/year	$260,000
C.O.D. surcharge	× 5%
C.O.D. surcharge/year	$13,000

While the first year cost is greater than the benefit derived from this program, the labor cost of purchase order approval enforcement can presumably be reduced in the second year, thereby rendering it cost/effective.

In summary, these cost/benefit examples can be used to develop real-life examples, especially in terms of the line items used. However, cost and benefit projections should use amounts derived from the specific situations of the reader, not from these examples.

REPORTS

A typical accounts payable system has many reports related to checks that have been cut or payments due. Under the revised system described in this chapter, additional reports are listed that can be used as control points or to reduce the number of payments made. For example, a report describing changes in payment dates on long-term contracts allows the controller to anticipate payment changes and avoid making incorrect payments that require time to correct after the fact. Also, a form used to order manufacturing supplies from key suppliers does not directly relate to accounts payable (since it is a purchasing function) but leads to having fewer purchase orders to match against invoices, which reduces the work-load of the payables staff. In short, a few extra reports and forms can improve the efficiency of the payables function.

One of the key control points in a revised accounts payable system is the purchase order. If a purchase order is stolen, it can be used to procure goods that will then be billed to the company. Though it is an after-the-fact control device, the following report is useful for spotting missing purchase order numbers; it lists the range of purchase order numbers used when printing a batch of pre-numbered purchase order forms. Any missing purchase order numbers are immediately apparent, since there will be a gap in the numbering of successive purchase order batches. A typical purchase order numbering report is shown in Exhibit 5.9. Note the gap in PO numbers between 3/9/09 and 3/10/09.

Of course, if the computer software contains the last purchase order number and prints the next one onto a blank purchase order, then this report is not needed.

Another report that is useful in making payments on long-term contracts is the one shown in Exhibit 5.10. It lists dates on which payment changes. This may seem minor, but considerable time isexpended whenever a payment is made in the incorrect amount, necessitating investigation, cancellation of the first check, and creation of a replacement check.

The form shown in Exhibit 5.11 is useful for ordering supplies from a local office supplies company while avoiding the usual purchasing transaction cycle involving a purchase requisition and purchase order. On this form, the user simply fills in the quantity of

Exhibit 5.9   Purchase Order Numbering Report

Date	Beginning No.	Ending No.
3/04/09	3400786	3400799
3/05/09	3400800	3400891
3/06/09	3400892	3400973
3/09/09	3400974	3401089
3/10/09	3401091	3401172
3/11/09	3401173	3401382
3/12/09	3401383	3401504

Exhibit 5.10 Contract Payment Dates Report

Supplier Code	Supplier	Change Date	Payment Amount	Payment Frequency
ARCOAL	Argentine Coal Corp.	3/3/09	$100.32/ton	Monthly
ARCOAL	Argentine Coal Corp.	3/3/10	102.81/ton	Monthly
ARCOAL	Argentine Coal Corp.	3/3/11	105.22/ton	Monthly
TEGAS	Texas Gas & Mineral	12/31/09	.52/cubic foot	Weekly
TEGAS	Texas Gas & Mineral	12/31/10	.54/cubic foot	Weekly
TEGAS	Texas Gas & Mineral	12/31/11	.57/cubic foot	Weekly
TEGAS	Texas Gas & Mineral	12/31/12	.62/cubic foot	Weekly

Exhibit 5.11 Office Supplies Order Form

SUPERFLOW TECHNOLOGIES
Office Supplies Order Form
Date _____
Deliver To _____

Description	Mfg. No.	Office Depot No.	Unit	Quantity
Binders				_____
Wilson Jones 8-½ × 11-½″ clear	70200	435-164	Each	_____
Envelopes				_____
Clasp envelope, 10 × 13	COR56	822-940	100/box	_____
Mailing envelope, 10 × 13, white	CO925	423-731	100/box	_____
#10 business envelope, white	CO196	804-724	100/box	_____
Folders				_____
Globe-Weiss letter file folder	321-1/3	475-814	100/box	_____
Globe-Weiss legal file folder	322-1/3	475-806	100/box	_____
Pendaflex letter hanging folder	4152-1/5	449-421	25/box	_____
Pendaflex legal hanging folder	4153-1/5	433-425	25/box	_____
Mailers				_____
10-1/2 × 16 air bubble mailer	18518	806-729	12 pack	_____
Paper				_____
Copier paper, letter	DC11	477-299	5K/case	_____

(Continued)

Exhibit 5.11   Continued

Description	Mfg. No.	Office Depot No.	Unit	Quantity
Copier paper, legal	DC14	479-253	5K/case	_____
Copier paper, 11″ × 17″	3R3761	345-603	500/ream	_____
Quadrille pad, letter size	76581	302-356	12 pack	_____
Paper clips				_____
#1 Regular	72380	808-881	1,000	_____
Jumbo	72580	808-907	1,000	_____
Small binder clips, 3/4″	72020	808-857	12 pack	_____
Medium binder clips, 1-3/4″	72050	808-865	12 pack	_____

each item needed and sends it to the office supply company. The description of each item as well as the part number used by the office supply company makes this an easy form for the supplier to process.

If a company pays suppliers based on the prices listed on the purchase order, then a key report in the process is a list of quantity and price disputes with suppliers. It can be used by the controller not only as a "memory jogger" of items to be resolved but also as a list of the potentially most troublesome suppliers. If this report consistently shows the same suppliers taking issue with the prices paid or the quantities used to calculate the total reimbursement, then it may be time to find a new supplier. The report shown in Exhibit 5.12 lists the prices

Exhibit 5.12   Supplier Disputes Report

Supplier	PO Number	PO Line Item	PO Price	Supplier Price	Price Variance	Supplier Quantity	Received Quantity	Qty Variance
Davis	92456	17	$12.32	$12.56	+.24	1,000	998	−2
Davis	93405	42	72.31	75.01	+2.70	500	459	−41
Montford	92155	03	88.33	90.11	+1.78	250	249	−1
Montford	93602	01	90.14	90.22	+.08	1,500	1,504	+4
Montford	94111	02	67.54	75.75	+8.21	100	92	−8
Thimbles	91002	01	12.57	15.62	+3.05	100	92	−8
Thimbles	93067	05	1.45	3.00	+1.55	50	46	−4

Exhibit 5.13   Amounts Due to Suppliers Based on Production Quantities

Finished Goods (F/G)	F/G Qty Made	Part Used To Make F/G	Part Qty	Extended Part Qty	Part Price	Extended Part Price
Wheelbarrow	32	Wheel	1	32	$3.50	$112.00
		Hand grips	2	64	1.22	78.08
		Legs	2	64	5.58	357.12
		Bin	1	32	8.73	279.36
Shovel	1,114	Scoop	1	1,114	4.82	5,369.48
		Handle	1	1,114	2.73	3,041.22
Shears	2,458	Hand Grips	2	4,916	1.22	5,997.52
		Shears	2	4,916	3.50	17,206.00
		Bolt assembly	1	2,458	.53	1,302.74

and quantities at issue in specific purchase orders, and is sorted by supplier.

A very useful evaluated receipts report is one that lists the amounts of finished goods in any time period, with payments to suppliers clearly indicated based on the amount of finished goods. A sample of such a report is shown in Exhibit 5.13. The report lists the quantity of finished goods produced on a given day, the quantity of specific parts contained in the finished goods, and the price to be paid to suppliers based on the quantities of finished goods completed. This report can be used as a detailed backup to payments in case suppliers complain about the amount of payments made. This report can only be used if there is one supplier for every part; with more than one supplier, this report cannot itemize which supplier to pay.

If an automated system is used to process employee expense transactions, then the controller should know about instances when employees are not providing receipts for expenses that exceed the company-mandated minimum expense level. If properly designed, the automated expense processing system should allow the employee to enter a reason for any missing receipts. This information can then be re-arranged into a report format for review. This report is very useful for spotting employees who are habitual offenders in terms of losing

Exhibit 5.14   Missing Expense Receipts Report

Employee	Expense Report Date	Expense Report Line Item	Line Item Desc.	Amount	Explanation
Barney, George	1/17/09	04	Airline	$542.00	Lost wallet w/receipt
Barney, George	1/17/09	05	Car	148.00	Lost wallet w/receipt
Chumley, Fred	1/24/09	02	Airline	1,042.00	Briefcase stolen
Chumley, Fred	1/24/09	03	Hotel	493.00	Briefcase stolen
Davies, Alice	1/17/09	01	Meal	82.31	Lost the receipt
Davies, Alice	1/17/09	07	Airline	951.00	Lost the receipt
Eddings, Percy	1/24/09	13	Car	232.00	Lost the receipt

receipts; additional education of those employees, as well as one-on-one discussions, is usually sufficient to correct the problem. A sample report is shown in Exhibit 5.14.

If an automated expense reporting system is used, then a key performance measure related to it is the turnaround time required to get a payment back to the employee for expenses submitted to the company. The report shown in Exhibit 5.15 lists the date on which an expense report was electronically submitted, the date the payment was made, and the time lapse between those two events. Also, since it is a major cause of delay in the payment of employees, the time required to obtain electronic approval from the supervisor is also listed.

In summary, the previous set of reports can be used for a variety of reasons—reducing the number of purchase orders requiring matching to invoices, spotting troublesome suppliers, listing employees who consistently cause problems on their expense reports, highlighting missing purchase order numbers, and so on. When used together, they

Exhibit 5.15   Expense Report Transaction Times

Employee No.	Employee	Filing Date	Payment Date	Time Lapse (days)	Approval Period (days)
000045	Bailey, Brad	5/3/09	5/10/09	7	4
000105	Davis, Charlie	5/1/09	5/21/09	20	18
000271	Ermine, Nancy	5/9/09	6/4/09	26	23
000521	Fingal, Mandy	5/2/09	5/15/09	13	2
000947	Gretz, Nimrod	5/8/09	5/12/09	4	1
001004	Smith, Herbert	5/1/09	5/8/09	7	1
002467	Yantz, Alfred	5/4/09	5/22/09	18	7

allow one to both reduce the accounts payable staff's workload and highlight payables areas that are causing additional work to be performed.

METRICS

The traditional measurement most frequently used to determine the performance of an accounts payable department is the number of supplier invoices processed per person. Most of the measurements in this section relate to changes made before the supplier invoice ever arrives at the company. Thus, most of these measurements highlight the volume of transactions about to reach the payables department.

NUMBER OF SUPPLIERS

The company must add up all the suppliers who sent invoices in the period just prior to implementing any improvements; this is the baseline number of suppliers. Then, as the number of suppliers is gradually reduced through the methods described earlier, this information can be reported on a trend line. To measure the current number of suppliers, simply add up the number of suppliers to whom payments were made in the past quarter. The purchasing department should compress this list on an ongoing basis, which allows the payables staff to consolidate its payments.

PERCENTAGE OF PAYMENTS WITH PURCHASE ORDERS

A very important measurement is the percentage of payments with related purchase orders. If a company is trying to move in the direction of paying suppliers from purchase orders instead of from supplier invoices, then this is the accounts payable measure to track. A small amount of programming will yield this measure—just summarize the total number of payments per month and divide that number into the number of payments that had a purchase order number in the payment record. The measure to strive for is 100% of all payments from a purchase order (with the exception of purchasing cards). To improve this metric, the purchasing and accounting managers should periodically review a listing of all payments that did not include a purchase order and incrementally implement additional controls that will ensure that purchase orders are issued the next time such transactions occur.

PERCENTAGE OF PURCHASE ORDERS BELOW A MINIMUM VALUE

A company should endeavor to purchase low-dollar items without purchase orders (usually with procurement cards or standard order forms), thereby reducing the workload of the purchasing, receiving, and payables departments. A company should create a threshold below which no purchase orders should be issued. Any such issuances can be reported by the computer system, and reviewed periodically by management.

PERCENTAGE OF TOTAL PAYMENTS FROM EVALUATED RECEIPTS

If a company is moving to payments based on completed production information, then a key performance measure is the percentage of total payments made from production records. When backed up by a detailed report that lists the suppliers not paid in this manner, this information can be used by management to target additional suppliers to be paid under this system.

SPEED OF EXPENSE REPORT TURNAROUND

When automated expense report processing is implemented, the time needed to manually complete the employee reimbursement transactions is minimal. Since employee-related costs are no longer of great concern, the best performance measure shifts into an entirely different area—the speed of expense report turnaround. This performance measure requires management to concentrate on getting payments to employees as quickly as possible, allowing them to pay their credit card companies on time, and thereby improving employee morale. The greatest problems found under this metric are the speed with which the accounting staff matches mailed-in receipts to the electronic report and (especially) the speed with which supervisors review and approve the electronic expense reports. The metric is calculated by using the filing date of the expense report as the start date, and the date of payment as the termination date. Then the average processing period is derived from all expense reports in the period, with detailed backup that provides information about specific expense reports that took an excessive amount of time to process.

NUMBER OF ACCOUNTS PAYABLE TRANSACTIONS

The last measure to track is the number of accounts payable transactions per payables employee. This measurement is the last one to review, since it will vary depending on the improvements being tracked by the other metrics already highlighted in this chapter.

COST PER PAYABLES TRANSACTION

Stated another way, a company can track the cost per accounts payable transaction. This is a more refined measure than the number of transactions per payables employee, since one can also focus on the related cost, which has a greater effect on the company's profits. Several studies have shown that a world-class company can process payables transactions at roughly one-quarter the cost of an average company, and the difference can be seen with this measurement.

In summary, most of the performance measures noted in this section are related to incremental and continuing improvements in the payables process. For example, the percentage of purchase orders issued below the minimum threshold requires management to continue to reduce the number of low-dollar purchase orders. Similarly, the expense report turnaround metric leads management to shrink the time required to get payments back to employees. By constantly reducing these transaction times, the accounting department will have more time available for other tasks.

SUMMARY

This chapter reviewed transactions required to process a payable in a typical company and then found a number of ways to reduce the time required: a single approval point (instead of three), payment from a single document (instead of three), increased use of procurement cards, simplified ordering systems, expense report automation, automatic approvals for small-dollar items, evaluated receipts, and more. There are a multitude of ways to streamline what is inherently a bloated and inefficient process.

Chapter 6

Cost Accounting

The cost accountant has traditionally dealt with a number of archaic variance analyses that no longer have much relevance. This chapter discusses measures that are more effective for managers, and also addresses such alternative systems as value stream management and throughput analysis.

CURRENT SYSTEM

The cost accountant usually creates a standard set of variances, investigates their causes, and initiates additional investigations if the first round of explanations is not adequate. This investigation is usually cursory, since the cost accountant is under considerable time pressure at month-end to deliver a reasonably plausible variance analysis, and the production personnel being questioned know this—if they give reasonable answers (e.g., excessive scrap due to some other department's problems, or extra overtime to achieve the monthly shipping goal), then the cost accountant will go away satisfied. After the month's results have been published, the cost accountant will make some attempt to verify the analysis, but the next period will be arriving in a few days, and then the cost accountant will abandon the last set of variances and concentrate on the variances for the following month. This cycle of activities does not allow for the conduct of in-depth analyses and does not contribute to fixing the underlying problems causing the variances. This section reviews the process in some detail.

A number of accounting functions must close before any cost accounting can begin. The accounts payable function must close, because invoices for additional inventory items must be added to the month-end inventory balance, and these bills can arrive several days after the end of the month. If the cost accountant were to begin work before this area closed, the book ending inventory balance could be too low. Next, the receivables function must close, because that closing process includes a review of the shipping log to ensure that all items shipped have been logged out of the system (which reduces the inventory balance). If the cost accountant were to begin work before this area closed, the book ending inventory balance would be too high, causing several variances to occur. Finally, the inventory function must close for the month. For companies that have either periodic physical inventory counts or perpetual inventories, this involves a reconciliation of the physical amounts to the book balances, requiring a further reconciliation of costs as well as quantities. Without this information, the cost accountant will have great difficulty in determining the causes of variances.

In those companies that close their inventory areas each month without a cross-check against physical balances, the cost accountant can begin work before the inventory area closes, because the information provided by the inventory function will be of no use—any materials variance explanations are guesses without concrete inventory information, so the cost accountant will be just as uninformed after the closing of the inventory function as before it closed. The example in Exhibit 6.1 illustrates this problem.

The cost accountant completes other work besides the materials variance analysis. Here are additional variances found in a traditional cost accounting system:

Exhibit 6.1   Example: Effect of Inaccurate Inventories on Variance Analysis

---

The controller of the XYZ Wholesale Products Company has nearly completed its January close and is calculating its ending inventory balance. Its beginning inventory is accurate at $914,000, because it was verified by the year-end physical count. The end-of-period inventory of $1,038,000 was derived by adding all purchases ($263,000) during the month to the beginning balance and then subtracting the items shipped ($139,000), according to the shipping log. The cost of the items shipped was determined by using the bills of materials for those items. The calculation was:

Beginning inventory	$914,000
+ Purchases	263,000
− Shipments	139,000
= Ending inventory	$1,038,000

The President of XYZ mentions to the controller that the upcoming year-end inventory had better not require the large (and disappointing) downward adjustment needed after last year's physical inventory. The controller responds that it is impossible to provide an ongoing materials variance analysis without a physical ending inventory each month. To prove it, he assigns the cost accountant, Mr. Forlorn, the task of providing a set of materials price and usage variances for January and documentation of how each variance was calculated without the benefit of a fully costed ending inventory. After a week of work, Mr. Forlorn, with an even longer face than usual, delivers the following report.

The materials usage variance was $25,000. Mr Forlorn had rooted through the scrap bin, dug out a number of discarded items, and individually costed them with the help of the purchasing staff. This had taken 20 hours. He recommended setting up a scrap reporting form on which, every time an item was thrown out, the quantity, part number, and reason for scrapping were to be entered. Also, a large number of items had been returned to suppliers; another form was needed to track these returns.

The materials price variance was $12,000. Mr. Forlorn had reviewed all large-dollar purchases received during the month and manually checked these costs against

(Continued)

the predetermined standards. This had also taken 20 hours. He noted that no one was checking actual prices against standards and that a database should be set up to report on the variance.

When the president and controller reviewed this information, they decided to maintain a database of purchase costs and to apply these costs to an ongoing set of perpetual inventory records. They determined that the effort required to do this equaled the effort required to research and create the materials usage and price variance reports.

- *Labor variances.* The major labor variances are for price and efficiency. The cost accountant calculates them as part of the period closing process. The information is calculated during the closing period and is given to production management, which is expected to shift the production staff to keep the cost of labor in line with the budget as well as to use the efficiency analysis to fix problems that created excessive labor usage in the previous period. In reality, the labor pool is relatively fixed, and the efficiency variance arrives too late to be of use to production management.

- *Overhead variances.* The major overhead variances are related to too much overhead being incurred (the spending variance), too much overhead being applied because too many base hours are being worked (the efficiency variance), and too much overhead being applied because additional units of product are being manufactured (the volume variance). This information is calculated during the closing period and is given to plant management and production management. Only the spending variance is of any use to production management, since it indicates too much money spent on specific overhead line items. The other overhead variances are an interesting academic exercise to review, but production management can do little to alter the remaining variances except change the level of production (which may be driven by a number of factors besides overhead cost control).

- *Closed job reviews.* One of the more useful roles of the cost accountant is the analysis of revenues and costs for completed jobs or production runs. This information can be calculated outside of the usual closing period and is given to production management. It contains the amount of costs incurred, and may do so in great detail, depending on the amount of work allowed for the review. The information is used to plan for the costs and prices of similar projects or production runs in the future.

In summary, the majority of cost accounting tasks are compressed into a small time frame during or shortly after the closing process. This results in hastily formulated variance reports that are so highly summarized that the information is of little use to production management, which needs detailed information. Also, the information is provided much too late for production management to solve issues; detailed variance information is required every day, not once a month. The cost accountant must report alternative information that is more useful to management, and do so on a more frequent basis.

REVISED SYSTEM

A revised cost accounting system should tackle the issue of speeding up the accounting process by addressing not only the tasks occurring during the closing process but also the relevance of the information transmitted to the operational staff. This allows the company to become more efficient (by speeding up the process) and more effective (by providing information that can be used to improve operations). This section first suggests cost accounting tasks that create better information and then examines how these affect the speed of accounting transactions.

If a cost accounting system is dysfunctional, it is probably caused by a flawed inventory record-keeping system. Inventory must be recorded under a *perpetual inventory system.* When the beginning or ending balances are incorrect, the cost of goods sold is incorrect. To avoid this problem, the cost accountant should review a large proportion of the inventory on a continuing basis and verify that the inventory has an accuracy level of at least 95%. This will help to reduce the materials quantity variance. Creating a perpetual inventory system is addressed in Chapter 4, Inventory.

Perhaps the greatest problem of all for the cost accountant is not having bills of material that are at least 98% accurate. An inaccurate bill can be used to purchase inaccurate quantities of materials or the wrong materials entirely. Furthermore, the cost accountant uses bills of material to cost products. Thus, if the bills are wrong, the costs are wrong. To avoid this problem, the cost accountant should be responsible for frequently *reviewing bills of material* with the engineering and production staffs to verify that they are accurate.

Management cannot accurately plan for future labor staffing if the labor routings associated with products are incorrect. This leads to labor price and efficiency variances, since incorrect staffing leads to having either over- or under-qualified staff (with associated pay rate variances) conducting either too little work (leading to negative efficiency variances) or too much work (costing the company in overtime expenses). To avoid this problem, the cost accountant should be responsible for frequently *reviewing labor routings* with the engineering and production staffs to verify that they are accurate.

If suppliers do not charge for materials at prices that are agreed upon in contracts, materials prices can spiral out of control. To avoid this problem, the cost accountant should compare actual materials prices to those noted in supplier agreements and report any variances to senior purchasing management.

Another cause of materials variances is the all-too-common write-off of large quantities of inventory during the year-end audit. This painful episode can be avoided by frequent *reviews of the inventory by the materials review board* (MRB), of which the cost accountant should be a member. The MRB is usually comprised of representatives of the purchasing, engineering, production, and accounting departments. The MRB's job is to identify slow-moving inventory and authorize its disposition in the most cost-effective manner. If the MRB is active, there should be few surprises in the year-end inventory that require a write-off.

Substantial changes in either production volumes or actual overhead costs may require a change in the overhead application rate. To spot these changes as early as possible, the cost accountant should monitor on a trend line the amount of overhead capitalized in each reporting period.

A reasonable question is whether all of the traditional period-end variance measures should be thrown out. These measures have been used for many years, and we should not be too quick to ignore the weight of so much tradition. One reason for continuing to calculate and report on the various overhead, materials, and labor variances is that they provide an overview of where costing problems are occurring. Thus, a quick glance at these traditional variances should pinpoint the localized area in which further investigation is warranted. Here are comments about how the variances can be used more productively:

- *Overhead.* Review the accounts that make up the overhead pool, and split the costs into fixed and variable elements. Usually, the bulk of the overhead costs will be such fixed costs as salaries, benefits, rent, depreciation, and utilities. Frequently, the only variable cost of any significance will be manufacturing supplies. If so, then it is easier to refer directly to the major variable costs as listed in the general ledger rather than compute the overhead spending variance. If production volumes are relatively stable, the overhead volume variance is unlikely to fluctuate much from period to period. Instead, track the difference between the actual overhead cost each period and the amount capitalized through the production process; any continuing difference between the two should be eliminated by a change in the overhead rate. Alternatively, if production volumes vary significantly from period to period, then report on the overhead volume variance.
- *Labor.* Many companies have a stable workforce that does not vary much from period to period. If so, consider labor costs to be fixed, rather than variable. The labor efficiency variance presumes that the staff is laid off the moment there is no work left to complete. In

reality, the staff is on site and being paid even if the workload drops substantially. Thus, the labor efficiency variance is not realistic when labor is a fixed cost. A more useful labor item to track is overtime as a percentage of total labor cost. Overtime information is useful for making long-term decisions about proper staffing levels.

• *Materials.* If there are highly accurate bills of material and an accurate inventory tracking system, then materials quantity variances are easily spotted without a high-level materials quantity variance analysis. The materials price variance is more deserving of publication. Any variation from long-term pricing contracts or purchase orders should be reported.

Thus, in many cases, the cost accountant should only review those variances that are most subject to short-term variation. Over the longer term, a comparison of the actual overhead cost to the amount capitalized, as well as periodic reviews of the average labor rate, will keep the cost accountant in touch with costs. However, these reduced levels of control are sufficient only if other day-to-day measures are carefully controlled, such as the accuracy of inventory, labor routings, and bills of material.

The cost accounting chore varies considerably if there is a *just-in-time (JIT) manufacturing system* in place. This system only produces enough product to meet customer needs, and emphasizes the reduction of cycle time, since employees can then produce quickly while still maintaining minimal levels of work-in-process. Since the collection of variance information during the production process increases cycle time, variance analysis is discouraged in a JIT system.

Without variance analysis, how does a JIT facility discover problems in the process? The key is the near-absence of work-in-process (WIP) inventory. A machine operator in a traditional assembly line process can continue to manufacture sub-standard parts without anyone discovering the problem, since the parts are loaded into a WIP queue and may not reach the next workstation for days. By the time the problem is discovered by the next downstream workstation operator, a large number of parts will have been produced, which must then be scrapped or reworked.

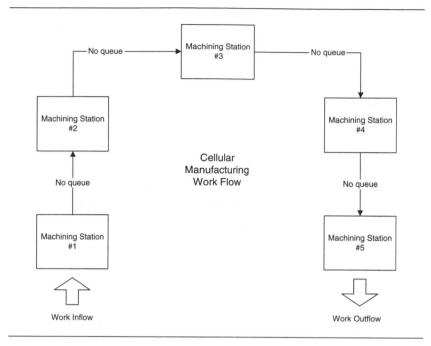

Exhibit 6.2 WIP Volumes in a JIT Environment

As shown in Exhibit 6.2, a JIT system forces the same machine operator to take the part to the next workstation and process it further (known as cellular manufacturing), at which time the defect will be discovered. Thus, under a cellular manufacturing system, only one part will be defective before the problem is fixed. If traditional variance reporting had been used in this environment, the cost accountant would have taken too long to collect the information, analyze it, and report it back to management; with no WIP to hide the problem, management would have already found and fixed the error.

There are entirely new ways of dealing with cost accounting that completely avoid the use of variances. One is *value stream management (VSM)*. A value stream includes all activities required to create value for a product or product line. These expenses begin with the designing and engineering cost of a product, the usual manufacturing costs, and post-shipment servicing. Value stream costing intersects many departments. This makes it difficult to accumulate costs, since many employees will be responsible for servicing multiple products. This type of costing works best when employees are clearly assigned to specific products. If costs can be accumulated in this manner, then a company has clear visibility into its profitability by product or product line. Under value stream management, the emphasis shifts away from variance analysis, toward daily operator-generated reporting that is acted upon at once by the value stream team. Also, there is a strong tendency under VSM to track costs at the product line level, rather than the product level. Many costs, such as trade shows, can be reasonably assigned at the product line level that would be inappropriately allocated at the individual product level. Thus, more costs can be considered direct costs at the product line level, thereby eliminating some overhead cost allocations and yielding a better picture of profitability at an aggregated level.

Another option is not a unique methodology, but rather a general *revolt against the use of overhead allocations.* Overhead is generally not under the control of the manager to whom it is being allocated, so there is little chance of overhead cost reduction, and a great deal of activity to have the cost reallocated to someone else. At some level, overhead is the direct responsibility of someone, so it is more useful to track these costs at the level of the responsible person, and not to burden anyone else with the information.

Finally, a completely different alternative to variance analysis is *bottleneck, or throughput analysis.* As explained in the author's *Throughput Accounting* book, this concept assumes that the key to profitability is maximizing the use of a company's bottleneck operation. If you focus all of the company's energies on making the bottleneck as efficient as possible, then profits will be maximized. Working on improving any operation other than the bottleneck is a waste of money, because sales are still constrained by the bottleneck. In fact, improving the efficiency of any operation than the bottleneck wastes resources, because those other operations will now increase their production, which will pile up in front of the bottleneck operation.

Under throughput analysis, it makes no sense to report on the efficiency of any operation than the bottleneck, since managers might use a high reported level of inefficiency to improve an operation that does not need improving, and should not be improved. Also, a traditional reporting system that focuses on reducing inventory levels might result in the removal of buffer inventory from the bottleneck. This is an incorrect result, since the bottleneck must be run at full speed at all times, so extra inventory acts as a buffer in front of it. Reducing inventory from this location can result in stockout conditions that shut down the bottleneck, thereby reducing profits. Thus, inventory turnover reporting can result in incorrect management actions.

A traditional cost reporting system might also show that a workstation is being underutilized, resulting in elimination of the excess capacity. However, this is incorrect in a throughput analysis, since there must be enough "sprint capacity" located upstream from the bottleneck to ensure that the system can rapidly produce sufficient replacement parts on short notice to keep the bottleneck fully operational at all times. Thus, eliminating sprint capacity reduces profits, even though a capacity analysis might suggest otherwise.

Labor efficiency also has little relevance in a bottleneck analysis. If a person is needed 10% of the time to run the bottleneck operation, then that is an excellent use of that person's time, because the bottleneck drives overall company profitability. However, a traditional labor efficiency variance would arrive at the appalling conclusion that the same person is wasting 90% of his time, probably resulting in an immediate termination. Again, variance analysis is harmful to overall profitability.

A final throughput concept is the amount of profit generated by each product that must use the bottleneck operation. If a product earns $10 of profit and requires one minute of bottleneck processing time, then this is a more valuable product than one that generates $20 of profit, but requires four minutes of bottleneck processing time. The first product generates $10 per minute, while the second only generates $5 per minute. Thus, the cost accountant should regularly report on this number for all products, so that management will know which products to emphasize through its pricing and promotions strategy.

In summary, the relevance of the information provided by the cost accountant is greatly improved by switching away from monthly variance reporting and into a review of the accuracy of such manufacturing data as the accuracy of inventory, bills of materials, and labor routings. Also, the cost accountant should advocate a switch to a JIT manufacturing system, so that the bulk of the variance analysis can be performed by the production crew instead of the cost accountant. Further possibilities are the use of value stream management and throughput analysis, which require completely different sets of management information.

CONTROL ISSUES

There are many control issues related to cost accounting. For example, if the cost accountant eliminated or reduced the level of examination of certain variances, a knowledgeable person could obtain company assets, knowing that the variance which would normally have flagged the theft was no longer being reviewed. This section contains examples of reduced controls, along with possible solutions to the resulting control problems.

* *Labor efficiency variance—embezzlement.* This variance shows the efficiency of the labor force versus the standard labor time expected to be used in production. The cost accountant conducts a minimal review of this variance, because the production staff is considered a fixed cost and will be paid even if they are not efficient. A production supervisor realizes that the variance is not being reviewed and continues to clock in a former employee. The paycheck is sent to the ex-employee by direct deposit, who splits it with the production supervisor.

    Solutions. The controller can ask the internal audit department to physically match employees to payroll records. Also, the controller can halt direct deposits occasionally and hand out checks directly to the staff, marking down recipients on a master list to ensure that no one receives more than one check. A variation on this method is to hand out small bonus checks in person, rather than halting the usual direct deposits for paychecks.

- *Labor efficiency variance—over-staffing.* Once again, the cost accountant conducts a minimal review of this variance. Realizing the lack of oversight, the production supervisor keeps more staff on hand than is required by the production schedule, so that the production schedule can still be met if sudden production increases are required. This costs the company, since the extra personnel are paid during slow periods, but it makes the production manager look top-notch when production deadlines must be met.

  Solutions. The company can pay the production supervisor a bonus incentive tied to the level of production staffing, so that the bonus declines if over-staffing occurs. Another option is to conduct just a high-level review, calculating labor as a percentage of material costs (or some other base that is closely tied to labor) and investigate large variances.

- *Labor rate variance—mix change.* The cost accountant conducts a minimal review of the labor cost per person, since this is considered a fixed cost. The production supervisor realizes this, and replaces all low-paid, inexperienced staff with highly-paid, skilled workers; this makes the production crew (and hence the supervisor) look very efficient, but is expensive for the company.

  Solutions. The production supervisor can receive a bonus based on the average wage received by the production staff, though this may push the supervisor in the opposite direction, where the staff is too *in*experienced. Another option is to conduct a high-level review and only investigate large variances. Yet another option is to calculate the average employee age from payroll records. On the assumption that older staff are more experienced and therefore better paid, any sudden jump in the average age can indicate a deliberate shift to higher-paid staff.

- *Materials usage variance.* This variance shows the difference between expected and actual materials usage. The company has restructured its manufacturing processes so that cellular manufacturing has replaced independent workstations. As a result, work-in-process levels have dropped significantly, allowing management to focus on scrap problems. After much work, scrap has been reduced to the point where the cost accountant has decided not to conduct detailed reviews of wasted materials. Workers in several cells become less careful in their work, resulting in increased levels of scrap. Realizing that the materials variance is no longer being reviewed, they throw away the scrapped materials and do not report the loss.

Solutions. Production employee bonuses that are tied to scrap levels will help to keep scrap down without any variance analysis by the cost accountant. If the company has accurate bills of material, the cost accountant can create a report that multiples the number of units shipped by the parts listed in each product's bill of materials, yielding a cost list of parts that should have been used during the period. When summarized, the total cost from this report can be compared to the actual materials cost for the period as a reasonableness check. If there is a notable variance, one can then compare the usage quantities on the report to the actual quantities used for each part. This allows for the highlighting of specific part usage, which can frequently be tied to the actions of specific production processes that use those parts.

- *Overhead efficiency variance.* This variance shows the difference between the standard overhead that would have been applied if labor had exactly matched the standard, and the overhead applied based on the actual amount of labor. The cost accountant has chosen not to track this variance. The amount of labor starts to rise, which leads to a similar increase in the amount of overhead charged to products.

  Solutions. The overhead efficiency variance is founded upon the belief that all overhead must be allocated using labor. An activity-based costing system uses multiple allocation bases, so that costs may be much more precisely assigned to projects. Thus, the use of activity-based costing nullifies the need to track the overhead efficiency variance.

• *Overhead volume variance.* This variance shows the difference between the standard overhead that would have been applied if the production volume had matched the budgeted volume, and the overhead applied based on the actual amount of production units. The cost accountant elects to ignore the overhead volume variance. Knowing this, the production supervisor decides to manufacture too much product, so that extra overhead will be absorbed, thereby creating a profit by capitalizing overhead into inventory.

Solutions. As usual, employee behavior can be modified by offering bonuses if certain criteria are met. In this case, the production supervisor will receive a bonus if production exactly matches demand. Another possibility is to review WIP and finished goods inventory levels to see if they are increasing. Higher WIP and finished goods inventory levels indicate that overhead is being capitalized into inventory.

A volume variance table is an easy way to quickly determine the expected gain or loss when overhead and production are at certain levels, all other variables being equal. An example of a volume variance table is shown in Exhibit 6.3. This information is useful for obtaining a high-level estimate of the overhead volume variance without being burdened by an excessive amount of variance analysis. The table shows a zero variance in its upper left-hand

Exhibit 6.3   Volume Variance Table

Overhead Rate	Production Level				
	$200,000	$225,000	$250,000	$275,000	$300,000
100%	$0	$10,000	$20,000	$30,000	$40,000
90	−10,000	0	10,000	20,000	30,000
80	−20,000	−10,000	0	10,000	20,000
70	−30,000	−20,000	−10,000	0	10,000
60	−40,000	−30,000	−20,000	−10,000	0

corner; this position uses the budgeted overhead and production volumes that were set at the beginning of the budgeting period. As production volumes rise above the budgeted level, the cost accountant can predict the amount of volume variance and recommend a new overhead rate that will eliminate the volume variance. For example, if the production level rises to $275,000, the overhead rate should be adjusted to 70% in order to reduce the volume variance to zero. This approach should stop production management from attempting to create profits by building inventories.

The following two variances are more critical as control points, and so should not be sidestepped:

• *Materials price variance.* This variance shows the difference between the actual and budgeted materials cost. This is the primary focus of the cost accountant, since it is the most variable item within the cost of goods sold. This area should not be subject to reduced review; on the contrary, the cost accountant may move to an even earlier step in the process and become involved in the cost of parts as they are engineered into a new product design (e.g., target costing). This action keeps costs out of the product before the first design is ever sent to the procurement department for sourcing with suppliers.

• *Overhead spending variance.* This variance shows the difference in actual overhead costs from the budgeted amount. The overhead spending variance is caused by costs being incurred in excess of those budgeted for the period. When overhead costs are applied to a product at a predetermined rate when the actual overhead rate should be higher, a spending variance occurs. This is not a variance that a cost accountant can overlook, because changes in overhead costs must be constantly monitored and promptly reported to management. Since overhead continues to take a larger portion of corporate costs, the cost accountant should be vigilant in watching over it.

Having been through all of these controls, let's review the situation from the perspective of throughput accounting. As noted in the last section, the only issue that matters in throughput analysis is to ensure that the bottleneck operation is always operating at 100% capacity. Therefore, all controls should be clustered around the goal of ensuring that 100% usage occurs. For example, there should be a trend line analysis of daily bottleneck usage. Another control is tracking of maintenance downtime for the bottleneck operation, for the same reason. Also, consider comparing the size of scheduled to actual production runs at the bottleneck, since it should always run the exact quantity needed; any quantity over that amount will sit idly in finished goods inventory, and so has wasted bottleneck time. Another possible control is to track all scrap that occurs downstream from the bottleneck operation. This scrap represents wasted bottleneck production time, so every effort should be made to eliminate this particular type of scrap. Any scrap occurring prior to the bottleneck is much less important. Further, create a report showing any time when the buffer inventory positioned in front of the bottleneck operation is reduced to zero, since this indicates that the bottleneck must have been shut down at that point. It is also useful to track the amount of sprint capacity at all key upstream work centers, to ensure that there is always enough to rapidly refill the inventory buffer.

None of the preceding throughput controls bear any resemblance to the traditional costing controls. The reason is that traditional controls emphasize the avoidance of costs, whereas throughput controls focus on increasing revenue. Thus, the controls have differing objectives.

In short, some variances can be ignored without harm to the company's overall cost control system by creating compensating controls and strengthening other existing controls. However, for an alternative focus on revenue generation, consider using the controls for a throughput-based system.

COST/BENEFIT ANALYSIS

It is difficult to construct a clear-cut statement of costs and benefits for the changes advocated in this chapter, but a number of justifications for the changes are explained here. The reader can use these justifications to construct a cost/benefit analysis; the analysis should include a range of values based on worst/expected/best cases for each item.

- *Cost of less review of the labor efficiency variance.* This is the cost of having an excessive amount of staff on hand in relation to the production level. This cost is difficult to determine, since proper staffing levels are difficult to maintain in the short term, and there is a cost associated with constantly laying off and recalling employees to exactly match the level of production.
- *Cost of less review of the labor rate variance.* This is the cost of having too many highly paid employees in the work force in relation to the skill level required to manufacture products. This cost is difficult to determine. A correct mix is difficult to maintain, since existing employees will become more experienced (and therefore more highly paid) with time.
- *Cost of less review of the materials usage variance.* This is the cost of too much scrapped material. This cost can be estimated based on current scrap levels and expected changes in that level when the variance review is decreased or eliminated.
- *Benefit of more review of materials prices.* This is the benefit of ensuring that suppliers meet their contracted pricing obligations. The benefit is estimated based on expected percentages of price overages that are discovered and corrected.
- *Benefit of reduced cost of data recording.* This is the benefit associated with not requiring employees to record variance data for labor and materials. The benefit is estimated based on the time previously needed to record the variance information.
- *Benefit of reduced cost of variance analysis.* This is the benefit associated with not requiring the cost accountant to spend time reviewing some variances. The benefit is estimated based on the reduced number of hours needed by the cost accountant.

A typical cost/benefit analysis for revising the cost accounting system is shown in Exhibit 6.4. It includes costs and benefits for

Exhibit 6.4 Cost/Benefit Analysis for a Revised Cost Accounting System

	Worst Case	Expected Case	Best Case
Cost of less review of labor efficiency (assumes efficiency reductions of 10%, 5%, and 3%)	(50,000)	(15,000)	(5,000)
Cost of less review of labor rate (assumes rate increases of 5%, 3%, and 1%)	(23,000)	(8,000)	(2,000)
Cost of less review of materials usage (assumes increased scrap of 19%, 5%, and 2%)	(62,000)	(11,000)	(3,000)
Benefit of more review of materials prices (assumes decreased prices of 1%, 4%, and 5%)	10,000	79,000	98,000
Benefit of reduced cost of data recording (assumes decreased labor costs of 2%, 3%, and 3.25%)	28,000	37,000	41,000
Benefit of reduced cost of variance analysis (assumes decreased cost accountant time of 40%, 40%, and 40%)	17,000	17,000	17,000
Net (Cost)/Benefit	(80,000)	99,000	146,000

three cases—worst, expected, and best. Some of these costs simply reduce the time requirements of personnel who will still be employed after the conversion to the revised costing system; thus, their full cost will still be paid by the company. An example is the reduced cost of variance analysis: The cost accountant will no longer be spending time on certain variance reviews, but will still be working for the company on new projects, so the true cost savings to the company will be zero.

In summary, it is difficult to construct a cost/benefit model for changes in variance reporting, but using a risk table based on worst, expected, and best case results allows one to construct a range of possible costs and benefits for changes to the cost accounting system.

REPORTS

Cost accounting reports vary considerably, depending on the types of variances being tracked. Under the traditional system, a number of efficiency, spending, and volume variances are aggregated into a single report, such as the one shown in Exhibit 6.5. This report seems clear enough—it shows how the variances are derived and how they all roll up into the total variance.

However, cost accounting is not about presenting neatly laid-out reports, but rather about revealing the reasons for costing problems. Thus, Exhibit 6.5 yields the following problems:

Exhibit 6.5   Traditional Cost Accounting Variance Report

Derivation of Total Variance	
Beginning cost of goods sold	$1,000,000
+ Purchases	300,000
− Shipments	150,000
− Ending inventory balance	825,000
= Cost of goods sold	325,000
− Cost of goods sold at standard costs	280,000
= Total variance	$45,000
Explanation of Total Variance	
Labor variance	
Labor rate variance	$3,000
Labor efficiency variance	1,500
Overhead variance	
Overhead spending variance	6,500
Overhead efficiency variance	4,500
Overhead volume variance	7,500
Materials variance	
Materials usage variance	9,000
Materials price variance	7,000
Explained variance	39,000
Unexplained variance	6,000
Total all variances	$45,000

- *Labor rate variance.* Corporate labor costs tend to be fixed in the short run, so a production manager cannot use the labor rate variance in any practical way.

- *Labor efficiency variance.* An efficiency variance is not of much use to a production manager, because there is no "drill down" to the next level of variance detail. For more detail, the cost accountant would need to review the accuracy of the labor routings used to measure the efficiency of the production staff and determine why the staff cannot match its efficiency goals. Too often, the problem is caused by inaccurate labor routings.

- *Overhead spending variance.* The production manager frequently has no control over some contents of the overhead account, so it is not of much use to know that overhead spending is greater than expected.

- *Overhead efficiency variance.* This variance is derived from the allocation base being either higher or lower than expected, resulting in unusual amounts of overhead being charged. The allocation base can be a variety of things—machine hours, production hours, square footage—which does not always relate directly to the overhead being charged. Thus, there may be no direct control over the allocation base that causes the variance.

- *Overhead volume variance.* This variance is derived from too many units being produced. Realistically, if production volumes increase beyond the expected range, then the overhead application rate should be reduced. Since the overhead rate can be changed at will, the variance has no particular value.

- *Materials usage variance.* This variance has some use, since it can signal the presence of incorrect bills of material, or incorrect inventory records, or incorrect picking transactions, or excessive scrap levels. However, the variance only indicates the presence of a number of possible issues.

- *Materials price variance.* This is a useful variance, since it can reveal actionable items. However, there must be sufficient additional detail for managers to correct pricing problems.

Exhibit 6.6 Improved Cost Accounting Variance Report

	This Month	Last Month	Last Year
Labor Information			
Average labor rate	$17.73	$17.71	$15.21
Average overtime percent	8%	12%	4%
Overhead Information			
Production supplies	$3,451	$3,200	$6,215
Supervisory salaries (average)	$82,400	$81,300	$71,000
Purchasing salaries (average)	$68,000	$68,000	$63,000
Production control salaries (average)	$52,000	$51,000	$47,000
Percent of overhead capitalized	92%	93%	97%
Materials Information			
Inventory accuracy percentage	94%	98%	97%
Bill of materials accuracy percentage	92%	99%	99%
Labor routing accuracy percentage	90%	98%	96%
Variance of actual purchased costs from targeted costs	102%	105%	107%
Working Capital Information			
Inventory turnover	5×	6×	8×

Clearly, the traditional variance report does little to assist in managing costs. The substantially revised format shown in Exhibit 6.6 provides significantly more useful information.

A few comments on the line items reported in Exhibit 6.6 are:

- *Average labor rate*. This rate gives the production manager a direct indication of changes in the cost of personnel. A more detailed report listing individual labor rates would be a reasonable supporting document.

- *Average overtime percentage*. This is a good indicator for when staffing levels should be changed, if overtime consistently exceeds an optimal level.

- *Production supplies.* This line item is totally under the control of the production manager. Any other fully controllable expenses can also be shown.

- *Percent of overhead capitalized.* If this amount varies much from 100% over an extended time period, then alter the overhead rate to match actual or projected overhead absorption.

- *Inventory accuracy.* If the accuracy measure drops below 95% in a material requirements planning (MRP) system, the system will produce inaccurate materials replenishment orders.

- *Bill of materials accuracy.* This may be the most important measurement of all, since it is used to derive purchasing and picking quantities. An accuracy level below 98% should trigger immediate corrective action.

- *Labor routing accuracy.* If the routing is inaccurate, it will indicate an incorrect crew type and staffing level to produce a product. Accuracy below 95% requires corrective action.

- *Variance of actual purchased costs from targeted costs.* This shows how closely a company is matching its targeted material costs. Further detail is always needed, showing which suppliers are not meeting parts pricing targets on specific items.

- *Inventory turnover.* If turnover drops, then the company is investing too much cash in inventory. This can be an indicator of obsolete inventory buildup, excessive finished goods in stock, or purchasing in excessive quantities. More detail must be provided to reveal the underlying problem.

Thus, a revised cost accounting system results in reports that differ substantially from more traditional costing reports. A revised system focuses on the continuing accuracy of key databases, rather than variances. It also reports on information that is at least one level closer to the basic problem than the information provided in a traditional system. For example, a revised system reveals the average percentage of overtime worked by person, whereas a traditional system will only report that the cost of labor has increased. These reporting changes

Exhibit 6.7   Throughput Contribution Report

Product Name	Price	Variable Cost	Throughput	Bottleneck Time Used (minutes)	Throughput Time per Unit
42″ Plasma TV	$3,100	$2,020	$1,080	20	$54.00
24″ Monitor	800	520	280	6	46.67
13″ B&W TV	120	70	50	2	25.00
30″ LCD TV	1,200	575	625	40	15.63
19″ LCD Monitor	290	230	60	14	4.29

allow report users to more easily find and correct cost-related problems more quickly than would be the case with a traditional reporting system.

The reporting used by a throughput-based system will not include variance analysis at all. Instead, there is a strong focus on the amount of throughput (price less variable expenses) per minute of bottleneck time, so that a company maximizes its overall profitability. An example is shown in Exhibit 6.7, where the dollars of throughput per minute of bottleneck time are shown for each product.

The throughput contribution report is sorted in declining order by throughput time per unit, so the products yielding the best return on bottleneck usage are listed at the top. In the exhibit, the 42-inch plasma television yields the best throughput per minute of bottleneck usage, at $54.00 per minute.

Another key aspect of throughput management is ensuring that there is an inventory buffer in front of the bottleneck operation that is of sufficient size to avoid any bottleneck work stoppage. The buffer management report shown in Exhibit 6.8 is designed to give details about problems that cause shrinkage in the size of the buffer (buffer penetration).

In summary, the cost accountant needs to confer with company management to determine what focus it wishes to place on its cost reporting—shall the emphasis be on cost reduction or revenue enhancement? A different set of reports are called for, depending upon management's focus.

Exhibit 6.8 Buffer Management Report

Date	Arrival Time Required	Actual Arrival Time	Originating Work Station	Cause of Delay
Sept. 11	9/11, 2 P.M.	9/12, 3 P.M.	Paint shop	Paint nozzles clogged
Sept. 14	9/14, 9 A.M.	9/16, 4 P.M.	Electrolysis	Power outage
Sept. 19	9/19, 10 A.M.	9/19, 4 P.M.	Electrolysis	Electrodes corroded
Sept. 19	9/19, 4 P.M.	9/25, 10 A.M.	Paint shop	Paint nozzles clogged
Sept. 23	9/23, 9 A.M.	9/24, 9 A.M.	Paint shop	Ran out of paint

METRICS

Exhibit 6.9 describes the formulas used to track the key information for a revised cost accounting system, as shown earlier in Exhibit 6.6. It does not include those variances used under an old-style costing system.

It is necessary to use a different set of ratios when evaluating the performance of a JIT manufacturing system. A JIT system operates on the principle that the facility should only receive enough supplier components to build parts, produce only enough parts to build the desired number of products, and only produce enough products to meet demand. In order to produce with the exact number of required components from suppliers, the receipts must be delivered to the company on time, in the right quantities, and with perfect quality (no defective components). In order to produce only enough parts to build the desired number of products, setup times must be minimized, work-in-process must be drastically reduced, and scrap must be carefully tracked. In short, the controller must devise data collection procedures for information that does not appear on the financial statements. The only ratio related to JIT that can be derived from the balance sheet is inventory turnover.

A throughput-based system calls for entirely different metrics. As noted in the Revised System section, a throughput-based system is solely concerned with maximizing the bottleneck operation. To that

Exhibit 6.9   Formulas for Key Elements of a Revised Cost Accounting System

Element	Formula	Comment
Average labor rate	(Total labor dollars expended in period)/ (Total hours worked in period)	Can be broken down by department to show changes in rate for each manager's area
Average overtime percent	(Total hours worked in period)/ (Total regular hours worked in period)	Can be broken down by department to show changes in percentage for each manager's area
Overhead information	—	Includes only costs that are controllable by manufacturing management. Usually includes supplies and salaries related to manufacturing and materials management.
Percent of overhead capitalized	(Total actual overhead cost)/ (Total capitalized overhead cost)	Can be shown as the dollar amount under- or over-absorbed
Inventory accuracy percentage	(Total parts counted – Items with incorrect locations, quantities, descriptions, or units of measure)/(Total parts counted)	—
Bill of materials accuracy percentage	(Total bills reviewed – Items with incorrect parts, quantities, or units of measure)/(Total bills reviewed)	—
Labor routing accuracy percentage	(Total routings reviewed – Items with incorrect times, machines used, or employee types)/(Total routings reviewed)	—

| Variance of actual purchased costs from targeted costs | (Total cost of purchased parts)/(Total cost of purchased parts as per contracts or purchase orders) | Can be shown by supplier to highlight suppliers who consistently overcharge for materials shipped. Can also be shown by number of parts with incorrect prices, rather than by total variation for all purchased parts. |
| Inventory turnover | (Cost of goods sold)/ (Average inventory) | Can be broken down into turnover for raw materials, work-in-process, and finished goods, so that slow turnover areas can be more easily seen |

end, key metrics include overall bottleneck utilization, as well as the ratio of maintenance downtime to operating time on the bottleneck (with an obvious goal of minimizing maintenance). Another key metric is the amount of scrap occurring downstream from the bottleneck, since this scrap represents a total waste of bottleneck time. It is also important to track attainment of the production schedule by the bottleneck operation, to ensure that the correct jobs are being completed.

In short, a variety of alternative measures are available that provide more relevant costing information to management than the traditional set of volume and efficiency variances.

SUMMARY

This chapter has described how the traditional cost accounting position no longer provides information to management that is current or relevant. A new set of costing information is available that could be more useful to management, from the perspectives of both cost reduction and revenue enhancement. The chapter also notes multiple new reporting formats for issuing this information to management, as well as a discussion of control issues regarding the abandonment of some costing variances.

Chapter 7

Payroll

In most companies, payroll processing requires a large staff of clerks to enter hours worked, as well as a variety of changes to employee pay rates and deductions. A fairly standard software module then calculates pay for employees, and paychecks are distributed. Even a cursory review of the function reveals that the data input part of the payroll process is by far its most inefficient part. While the pay distribution function can certainly be improved as well, the bulk of this chapter addresses how to improve the data entry portion of payroll processing, so that fewer employees are required, data entry errors are decreased, and the payroll can be processed more rapidly than before.

CURRENT SYSTEM

This section breaks the payroll processing function into three steps and describes the processes and controls used in each.

PAYROLL DATA INPUT

The payroll data input step enters information from a variety of sources into the payroll software for later calculation. The information is typically entered by a team of payroll clerks. The most common types of information entered are as follows:

- *Employee hours worked.* This can be broken out by regular and overtime hours, as well as by piece rate. The system may also require separate identification of vacation or sick time. Also, depending on company policy, hourly employees may not be paid for various fractions of an hour if they arrive a few minutes late for work. To make matters even more complicated, some companies that charge costs to specific jobs must also assign hours to specific job numbers—this information is passed to the job costing system for summarization into total costs incurred for individual jobs. In short, the most time consuming and error-prone payroll data entry task is the one involving the entry of hours worked.

- *Bonuses and commissions.* The accounting staff calculates commissions based on such variables as a percentage of revenues, splits with other salespeople, splits with sales supervisors, cumulative year-to-date bonuses, regional premiums, and percentages of gross margins on products sold. Commissions may be paid based on cash received, revenue booked, or cash received less bad debts. Bonuses can be based on similar measures. In short, these calculations can be difficult to track and summarize, and the number of methods used increases with the ingenuity of the management team in coming up with new bonus and commission payment schemes.

- *Benefit deductions.* The typical company offers a broad range of dental, health, life insurance, and other benefit plans, each of which has a deductible that must be removed from employee paychecks. Since deductibles may change based on which set of benefit options is chosen, this can be a difficult deduction to track. Also, the paperwork is typically filled out by the employee and passed to the human resources department, which in turn gives the deduction information to the payroll clerks for entry into the computer. Since there are a number of people involved in the process, it is easy to enter an incorrect deduction amount.

- *Tax deductions.* Employees request changes to their deduction amounts from time to time for both federal and state withholdings. These changes may be a percentage of gross earnings or a flat amount. In either case, the payroll clerk makes the entry based on signed paperwork filled out and submitted by employees.

- *Pension plan deductions.* Some pension plans, such as the 401(k), require the employee to contribute either a percentage of gross earnings or a flat amount to a fund, to which the employer also contributes. Though some plans allow only biannual changes, the majority are more frequent. This means that the payroll staff may enter a large number of pension contribution deduction changes per year.

- *Garnishment deductions.* Companies are required by law to deduct specific amounts from an employee's paycheck to cover court-ordered payments resulting from a legal settlement. This requires not only the initial entry by the payroll clerk of the garnishment amount, but also tracking the recurring garnishment deduction until the deduction terminates.

- *Distributions to bank accounts.* If a company uses direct deposit to send payroll payments directly into employee bank accounts, then the payroll staff must enter the bank number and account number into the payroll system. In addition, employees may request distributions to multiple accounts, so that a single paycheck may be sent to a checking account and a savings account. All these accounts require a setup entry, and possibly a pre-note transaction as well, to ensure that the distribution will work as planned.

- *Shift differentials.* A company may pay its employees an additional amount per hour for working the second or third shifts, since these are less desirable times to work. The payroll staff must enter these shift differential premiums into the payroll system every time an employee changes shifts.

- *Personal item deductions.* Some companies allow their employees to purchase items through the company, since it can command reduced prices on a variety of products. The employees then pay the company through payroll deductions. If the item purchased is quite expensive, the deductions may span a number of payroll cycles. The payroll staff must track these deductions until the full amount has been paid.

- *Pay rate changes.* With supervisory approval, the payroll staff must enter pay changes for staff members. These can be more frequent than once a year, since new employees frequently receive a review after 90 days with the company.

- *New employee additions and departing employee deletions.* With proper supervisory approval, new employees must be added to the database. These entries may require such additional information as emergency contact names and phone numbers, social security numbers, mailing addresses, start dates, and initial pay rates. Also with proper supervisory approval, the payroll staff may delete departing employees from the payroll system, possibly with additional deductions for any liabilities owed to the company. In both cases, the payroll staff must calculate partial payments for partial periods worked, since employees do not always conveniently begin or stop work at the beginning or end of the company's payroll cycle.

Controls are needed when inputting payroll information. In particular, pay changes, new hires, and employee deletions are not input without a signed authorization from management. Also, hourly employee time cards are usually reviewed in advance by the supervisor of the hourly employees for any irregularities and signed before being passed on to the payroll clerk. Bonuses and commission calculations also are normally reviewed by a supervisor prior to payment. In general, any change to an employee's gross pay requires advance approval by management. However, pay deductions only require the approval of the employee. This is a key control issue that allows a company to automate key features of the payroll data input step.

PROCESSING

The process step takes all the information entered in the previous step to calculate gross wages from hours worked, commissions, and bonuses. It then deducts legally allowable items, such as 401(k) pension plan deductions, before calculating a variety of taxes. The system then deducts all remaining items to arrive at the net pay for each employee.

The processing step is nearly always carried out by payroll software or service providers. Only the smallest companies manually calculate payroll. A service provider likely has extensive automated controls, as will a payroll software package.

For companies with subsidiaries, it is possible that their payroll systems are not integrated, so there are multiple payroll systems for various subsets of employees. This can cause problems when employees rotate among subsidiaries, since the time of their tenure (used for pension vesting and vacation accruals) must be shifted to the payroll system of the subsidiary to which they are moving. Controls are needed to shut down an employee in one system and start him up in the succeeding system.

Controls over the processing step primarily relate to access to the program. Thus, the control is simple—access is restricted to specific user identification codes and passwords. From the perspective of the computer systems manager, it is wise to retain multiple off-site copies of the payroll database and application software in case of damage to the primary computer storage facility.

DISTRIBUTION OF PAYMENTS

The distribution of payments step involves printing checks, signing them, and distributing the checks to employees as follows:

1. *Set up printer.* Remove check stock from a secure location, and set it up on a printer. If used, also remove a signature plate from secure storage and set it up for use in the printer. The check stock and signature plate should be locked in separate locations, which makes it more difficult to steal them.
2. *Print checks.* This step may involve loading the first check number on the check stock into the software. Once completed, the payroll

staff should update a check number log, showing which checks have been used. This is useful for identifying missing checks.

3. *Sign checks.* Supervisors who sign checks can determine, based on their knowledge of who works for the company, if any checks are being cut for people who no longer work there, or for amounts that are clearly excessive. If the company has grown so large that the signer cannot know all employees, then the control becomes much less effective.

4. *Enclose checks.* The accounting staff seals the checks in envelopes, with the employee name appearing through a window; alternatively, and as an additional step, the staff can add labels to each envelope.

5. *Distribute checks.* An employee designated for the task (sometimes called a paymaster) distributes all checks into the hands of the employees to whom the checks are addressed. Checks should only be given to the employees to whom the checks are addressed. By enforcing this control, any checks created for nonexistent employees will never be issued. A final control over paycheck disbursement is that the paymaster not be involved in payroll processing. This control keeps a payroll clerk from creating a check and then disbursing it to himself.

Clearly, the traditional payroll process requires a great deal of paperwork movement among employees. The move and wait times thus introduced into the process greatly slow it down, and the potential exists for the loss or incorrect handling of the paperwork. A payroll value-added analysis is shown in Exhibit 7.1. A value-added item is considered to be one that brings the payroll transaction closer to conclusion. The time estimates in the table are based on a single working day of activity.

Exhibit 7.2 shows that only 16% of the steps bring the payroll transaction closer to conclusion; all other steps are related to moving paperwork from person to person or re-entering information that has already been completed by the employee. In terms of time required,

Exhibit 7.1 Payroll Value-Added Analysis

Step	Activity	Time Required (Minutes)	Type of Activity
1	Employee enters time worked on time card	5	Non-value-added
2	Payroll clerk moves to time card storage rack	1	Move
3	Payroll clerk removes time card from storage rack	1	Non-value-added
4	Payroll clerk returns to office	1	Move
5	Payroll clerk reviews time card for overtime	2	Non-value-added
6	Payroll clerk reviews time card for absences	2	Non-value-added
7	Payroll clerk reviews time card for missing time entries	2	Non-value-added
8	Payroll clerk enters hours worked into payroll software	5	Non-value-added
[Payroll clerk receives benefits deduction information from human resources department]			
9	Payroll clerk enters benefits deduction changes into payroll software	2	Non-value-added
[Payroll clerk receives tax deduction information from employees]			
10	Payroll clerk enters tax deduction changes into payroll software	2	Non-value-added
11	Software calculates amount payable to employee	0	Value-added
12	Clerk prints payroll checks	3	Value-added
13	Clerk brings checks to authorized check signer	1	Move
14	Authorized signer reviews and signs checks	1	Non-value-added
15	Clerk stuffs checks into envelopes	1	Non-value-added
16	Clerk brings checks to paymaster	1	Non-value-added
17	Paymaster delivers checks to employees	5	Value-added
18	Paymaster brings undistributed checks to accounting department	1	Move
19	Clerk stores undistributed checks in a safe	1	Non-value-added

Exhibit 7.2   Summary of Payroll Value-Added Analysis

Type of Activity	Number of Activities	Percentage Distribution	Number of Hours	Percentage Distribution
Value-added	3	16%	0.13	22%
Wait	0	0	0.00	0%
Move	4	21%	0.05	8%
Non-value-added	12	63%	0.42	70%
Total	19	100%	0.60	100%

the value-added steps can be concluded in eight minutes, whereas the non-value-added and move portions of the transaction take up more than triple that amount of time. Thus, the actions needed to pay an employee are a small proportion of the total process.

In summary, a large amount of time is required to input wage and deduction information into the payroll software. These entries are subject to input errors, with resulting employee dissatisfaction. The payroll processing is usually automated, with few processing problems, while the paycheck-printing process is also relatively free of problems. The next section discusses ways to reduce the work load of the payroll staff in entering payroll information, while at the same time reducing the number of entry errors and increasing the transaction speed of the entire process.

REVISED SYSTEM

This section discusses how to speed up the payroll process while cutting down on data entry labor and errors, resulting in greatly reduced costs.

If a company requires its employees to track an inordinate amount of information, then this requires not only more data entry by someone, but also an increased risk of incorrect data entry. Employees can also become frustrated, and ignore the various types of information that they are asked to enter. Besides charging time to specific jobs, employees may also be asked to record it as training, jury duty, holidays, vacation, sick time, bereavement leave, and family leave. In the manufacturing area, the number of standard pay codes can be over-whelming, including codes for receiving, putaway, picking, cross-docking, case breakdown, and cycle counting—and that is just the warehouse. The trouble with having a broad range of pay codes is that the resulting data does not yield any improvement in company opera-tions. Instead, once time is compiled initially within each pay code, the cost accountant will find that the proportions of time spent on each activity do not change very much over time. Thus, the extra labor re-quired to charge time to them only affirms that activities are usually in a steady state and do not require much monitoring.

A better approach is to have as many employees as possible *charge their time to a single pay code* that needs to be altered only on an excep-tion basis. If the cost accounting staff requires additional information, it can conduct a special one-time study and collect the information itself. An even better option is to use *exception time reporting*, where the payroll system automatically assumes a default number of hours for each em-ployee, which are only changed if an employee works a short week, over-time, or charges time to some special pay code, such as vacation time.

One of the problems with a traditional payroll system is the number of people who handle deduction information before it finally enters the payroll system. For example, an employee selects a medical plan; the human resources department calculates the employee-paid portion of the plan and forwards this amount to the payroll department, which enters the information. The employee receives a remittance advice that lists the amount of the deduction, and wants to change to a less expen-sive alternative. This sets off another round of changes. Not only does this approach involve the participation of too many people (any of whom could pass along incorrect information) but it may take too long to enter the information, especially if the deductions are being sent from an outlying location to a centralized payroll system. This is of considerable concern to a company, which grants a benefit to an em-ployee as soon as the employee signs the paperwork but which cannot enter the related deduction information into the payroll system quickly enough to recoup part of the benefit cost. Of course, the payroll staff can always make a double deduction in the following pay period to make up for the missing deduction from the prior period, but this requires additional time to make the manual entry.

A company can set up *employee self service*, so that employees can alter their own deductions. Alternatively, the same type of portal is offered by some payroll outsourcing providers. Since not all employees have access to a computer terminal, a company can provide access to a free-standing computer kiosk that allows access to the self-service system. A flowchart showing direct entry of deduction changes by employees is shown in Exhibit 7.3.

Employees enter their employee numbers and personal identification numbers (PINs) to gain access to the self-service system. To keep employees from being surprised at the net amount of their paychecks when paid, the system immediately calculates the amount of net pay left and presents this amount to the employee. If the net amount is acceptable, then the employee allows the change. Pension deductions operate in a similar manner. Benefits, however, are slightly different. Most companies have a variety of benefit options, so the employee may need to select a combination of several dental, medical, disability, or life insurance plan options, each of which has a different deduction amount. The system should communicate the effect on net pay to the employee as it is selected, so the employee can determine the cost of individual benefits, rather than only the cost of all the benefits together. Also, the company should provide employee training, so they will be more comfortable with using the system. In short, direct access to a self-service module allows employees to enter deduction information themselves with immediate feedback, which eliminates errors and keeps the payroll data entry staff from having to enter any deduction information at all.

Employee self-service can be extended considerably beyond the management of deductions. In addition, the system can allow employees to alter their addresses, tax withholdings, and direct deposit information. By giving employees this extra level of access, the payroll department will find that the number of manual entries it is called upon to make will decline precipitously. Systems providing this level

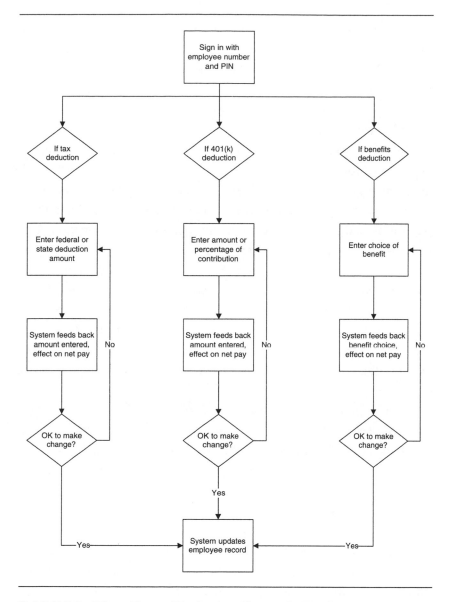

Exhibit 7.3   Direct Entry of Deduction Changes by Employees

of access are now commonly provided by payroll outsourcing firms, and are also available through higher-end accounting systems.

The concept of employee self-service can also be extended to company managers. This group can access a different *manager self-service software module or Internet portal* in order to review and approve the time entries of their employees, shift differential payments, overtime, and wage rate changes. This function is not so common as an employee self-service system, but is gradually being rolled out by the makers of more complex payroll software, as well as by the larger payroll outsourcers.

Another problem with the payroll process is having to make deduction entries for those employees who have purchased items through the company and who are now compensating it for the amount of the purchase. The payroll staff must enter the amount of the deduction per pay period and track the deductions through to their termination date, when the deduction is removed from the system. This process increases the payroll workload, not to mention the general ledger clerk, who must make a journal entry to account for the repayment. Since buying items for employees does not fit into the mission statement of any company, the best solution is to *create a policy of not allowing company purchases of merchandise on behalf of employees.* By removing the need for the deduction, the deduction-related work disappears.

Another problem is the sheer volume of benefit deductions. A company can set up deductions for employee medical, dental, life, and supplemental life insurance, flexible spending account deductions for medical insurance or child care payments, as well as 401(k) deductions and 401(k) loan deductions. If there are many employees and many deduction types, the payroll staff can be snowed under at payroll processing time by the volume of deduction changes. There are several ways to address this problem. One is to *eliminate the employee-paid portion of some types of insurance.* For example, if the employee-paid portion of dental insurance is only $2 per paycheck, would it be more worthwhile for the company to eliminate this deduction entirely, and instead shift the deduction to some other type of insurance? Another alternative is to *eliminate certain types of benefits*, such as supplemental life insurance, in order to eliminate the related deductions. Yet another option is to *create a policy that limits employee changes to benefit plans*, so they can only make a small number of alterations per year. A very good alternative is to *create a benefit package* for all employees that requires just a single deduction of the same amount for everyone; employees can then choose the exact amount of each type of benefit they want within the boundaries of the benefit package, without altering the amount of the underlying deduction. This last alternative has the unique advantage of consolidating all deductions into a single item, which is much simpler to administer. Any of these approaches to the problem will reduce the volume of deduction changes.

Another problem is the amount of work required to calculate commissions. These calculations can be so complex that it is too expensive to generate the program code needed to automate the process, resulting in manual calculations by the accounting staff. This problem is exacerbated if there are a large number of invoices to summarize by sales territory for each salesperson's commission calculation. The solution is a difficult one for the sales manager, who must adopt a *simpler commission structure*. The simplified structure is then automated, so the computer system generates a list of commissions payable without any input from the accounting staff. The sales manager does not like this, because the complexity of the typical commission system is designed to foster a certain type of behavior by the sales staff, and a change to that system may not result in optimal sales. However, by educating the sales manager about the time and effort required for commission calculations, there may be some hope of success.

From the perspective of the accounting department, the ideal commission structure is one that can be summarized directly from a standard accounting system report, followed in attractiveness by one requiring only a minimum of programming effort. In declining order of desirability, here are some commission calculation strategies:

1. *Percentage of invoiced dollars.* Many accounting packages can calculate this commission in a standard report. The typical accounting system already adds a salesperson code to each invoice, and should accept a standard commission percentage.

2. *Percentage of cash received.* Not so many accounting packages can calculate commissions that are only paid upon cash receipt. This is a

more difficult calculation process, since customers may not pay in full for each invoice, or make multiple incremental payments.

3. *Percentage of invoices issued, less bad debts.* If a company issues a credit memo for each bad debt, then there must be a procedure to assign a salesperson code to each credit memo, so that the related negative commission can be deducted from a salesperson's next commission calculation.

4. *Any preceding method, plus splits.* This is a more difficult programming task than may be immediately apparent, especially if combined with payments based on a percentage of cash received. This is because there must be room in the invoice file for multiple salesperson identification codes, as well as multiple commission percentage codes, which requires a change in the structure of some databases. In the case of commission payments based on cash received, it also means that the computer system must track the differing commission payments still owed to multiple salespeople on invoices that have only been partially paid.

5. *Any preceding method, plus different commission rates depending on gross margins.* The most difficult task of all is creating fields in the database for the gross margin on each product sold, as well as a separate table that issues a different commission rate based on the amount of the gross margin. To complicate matters further, the gross margin may be tied to actual costs, so that the gross margin may fluctuate over time. To make matters even worse, the commission table may have to store older commission rates in case salespeople are compensated based on commission rates existing at the time the product was sold, rather than the amount in effect when cash was received in payment for the invoice.

The highest-volume item involved in the payroll data entry process is entering hours worked for those employees who are paid on an hourly basis. This information is usually translated from time cards and manually entered into the payroll software by the payroll staff. Since the hours worked must be manually transferred, there is an enhanced probability of errors being made by the payroll department.

When errors are discovered, the payroll staff must bring in the questioned time card from storage, recalculate the time, correct the error, and file away the time card again—a considerable waste of time. To improve the situation, the company can install a *computerized time clock*. With this system, each employee is given an employee card with an identifying bar code or magnetic stripe on the back. When an employee arrives, leaves, or takes a break, he runs the card through a scanner, which automatically identifies the employee and records the time in a computer file. If the scanner is biometric, it will scan the outline of a person's hand. The scanner can also be used in association with a keypad to punch in the time worked on specific jobs. The scanner is linked to a computer that stores the time information and can be linked with an interface to the company's payroll software, so that all employee time information is uploaded for payroll processing without any manual data entry. A variation on this approach is the *Internet-based time clock*, where employees can log into a company Web site to enter their time information. This approach is particularly effective when a company's work force is widely distributed, as would be the case with a field sales force or consultants. There are even third-party services available that allow employees to *record their time from cell phones*, which transmit formatted text messages to a central timekeeping database. While a computerized time clock eliminates data entry, the payroll staff should still review an edit report for any obvious errors, prior to calculating payroll.

A computerized time clock has other benefits. For example, it can reject a scan if an employee has arrived late for work and will only allow the entry if a supervisor punches in an authorization code; this keeps management apprised of late arrivals. Also, the system can automatically enter an employee's missing time entry if she forgets to clock out. Another feature is not allowing time entries for shifts in which an employee is not supposed to be working, thereby keeping her from punching in for the wrong shift and collecting a pay differential for the shift. Finally, management can print a variety of reports from the system, such as a listing of who has not shown up for work, who is chronically late, and who forgets to clock in or out. In short, there are a variety of reasons why a computerized time clock can improve the overall payroll system.

A computerized time clock is relatively expensive, so a parsimonious manager might be tempted to buy too few of them. If so, employees will wait in long queues when it is time for them to clock in or out. This certainly does not improve employee acceptance of the new system! Instead, be sure to follow the manufacturer's recommendations for the number of clocks per employee, and also have one installed near every major exit from the building. The extra clocks will eliminate any hassle for the employees. Also, if employees are also expected to punch in and out of specific jobs, this may require a large number of scans during the day, and multiple trips to the time clock. This can require a startling amount of wasted time to walk to and from the nearest scanner, so be sure to liberally sprinkle extra clocks throughout the facility.

Another problem is the presence of multiple unlinked databases that are needed to process payroll. For example, the payroll database contains an employee's name, address, social security number, deductions, and current pay rate. The benefits database (which is manually compiled in many smaller companies) also lists each employee's benefit selections. Further, the human resources database contains such additional information as employee contact information, pay history, job classification, and history of work-related accidents. When maintained separately, there is a strong likelihood of varying addresses, deductions, and pay amounts being stored in each one, since not all of the information is updated when a change is made—to do that, a different person must enter the change in each database. The obvious solution is to *combine the three databases into one*. By doing so, any information change requires a single database update, and eliminates the time otherwise needed to reconcile the information in each database.

If all of the payroll-related databases have been consolidated, then it is also possible to *link payroll changes to employee events*. There are many changes in the various databases that must be made when key employee events occur, some of which are not likely to take place unless the computer system can automatically generate reminder messages to employees in the payroll and human resources areas. Here are some of the reminders that can be issued by a properly configured and consolidated computer system:

- Issue a COBRA form to any employee who has just left the company.
- Issue a pension eligibility notification to an employee, as well as the pension manager, as soon as an employee reaches the age where this action is triggered.
- Notify the human resources staff as soon as an employee has passed the probationary period, so that the person can be added to the various company benefit plans.
- Upon hiring, the purchasing department is notified that business cards must be ordered for the new employee.
- Upon notification of an employee address change, notifications will be sent to all suppliers who must update their files, such as the medical and pension plan administrators.
- Upon notification of marriage, an employee should be sent a new W-4 form to complete for tax withholding purposes.

These notification features are available in enterprise resources planning (ERP) software packages. In other cases, they must be custom-designed into existing applications.

Along similar lines, a company should *reduce the number of payroll systems*. If a company has grown by acquisition, it may have dozens of payroll systems. If top management demands consolidated payroll reports, it may be quite difficult to accumulate this information from a multitude of systems. This centralization process can be a lengthy one, and will require the centralization of the various payroll staffs.

Turning to the output of information from the payroll process, a company creates a paycheck for each employee. This creates a problem for those employees who are not on hand on payday to receive their checks. This is also a problem for centralized payroll systems, since the company must use an expensive overnight delivery service to send paychecks to outlying locations in time for payday. A cleaner approach is *using direct deposit or pay cards to pay employees.* With either method, pay amounts are automatically deposited in employee accounts on payday. This eliminates the need for a manual distribution of paychecks and allows employees to receive their pay even when they are not on hand to receive a paycheck on payday. A pay card is used instead of direct deposit when an employee does not have a bank account; instead, they withdraw cash as needed from an ATM.

A company will experience a number of additional benefits from either form of electronic payment. It will avoid all stop payment fees associated with lost checks, as well as all risk of escheatment, where an uncashed check eventually must be remitted to the local state government as unclaimed property. In addition, employees have no need to leave work to cash checks, so the company will benefit from more employee work hours. To further reduce employee travel time, a company can even install an ATM on its premises, so that pay card holders can withdraw cash without leaving the building.

It is still a good idea to give employees a remittance advice that lists their gross pay, deductions, and net pay, but as long as the money reaches employees on time, there is no need for the remittance advice to be given to employees at the same time—instead, less expensive postal service rates can be used to send remittance advices to outlying company locations. An alternative is *sending an email message to employees, as a notice that their remittance advices can be downloaded from a secure Web site.* This feature is offered by many payroll outsourcing companies. It eliminates the cost of paper and postage when sending out remittances, but is only good for those employees who have access to email.

A final issue that slows down the payroll staff is responding to inquiries by employees about the amount of vacation and sick time they have remaining. A simple way to avoid it is including the accrual amount on the remittance advice for each paycheck. This is quite effective for weekly pay periods, but less so if the pay period is monthly, since accruals may change considerably during the month. In the later case, the company can also provide the information through a secure Web site. The net effect is that more time is available to the payroll staff for other activities.

Most of the preceding recommendations are targeted at the complete avoidance of data entry by the payroll department. However, this is not always possible. If so, the payroll manager can adopt the following changes to reduce the impact of payroll data entry on her department:

- *Error handling system.* The payroll department is subject to the basic truth that payroll errors are painfully obvious, so that even a small number of errors appear magnified in the eyes of the work-force. To gradually whittle down the number of these errors, the department manager should draw up a formal list of errors each month and go over them with the payroll staff. By creating a *formal system for addressing transaction errors* and finding ways to avoid them in the future, there is a good chance that errors will decline over time. It is also useful to track error types on a trend line, to see which errors keep recurring. These are the ones to which the most attention should be directed.

- *Transaction cutoff.* Payroll transactions tend to surge into the payroll department just before processing of the latest payroll cycle begins. To keep the payroll staff from being buried by this avalanche, enforce a cutoff date by which all payroll changes can be submitted to the department. This allows the department to avoid any last-minute data entry crunch that would otherwise at least cause overtime, and probably a higher-than-usual incidence of errors.

- *Reduce payroll cycles.* A payroll department may be occupied with a multitude of payroll cycles, such as a weekly one for hourly employees and a biweekly one for salaried staff. A better option is to consolidate all payroll cycles into a single, company-wide cycle. In addition, lengthen the payroll cycles to eliminate all weekly cycles. By doing so, the payroll staff can vastly reduce the amount of its payroll processing effort.

- *Minimize off-cycle payrolls.* There are occasions when managers demand an off-cycle payroll, perhaps to correct a payroll error or issue special bonus payments. Every time this happens, the payroll staff must spend considerable additional time reviewing open transactions, summarizing hours worked, entering data, and so on. The department can reduce the incidence of these special payroll cycles by imposing a hefty interdepartmental fee to run one, or to require extremely high-level manager approval. If the normal cycle is semi-monthly, then the payroll manager can also use the excuse that the next regular payroll is not so far in the future that the special payment cannot wait for it.

In summary, a variety of techniques are available that will increase the speed of the payroll process while reducing the number of payroll transaction errors. By automating data collection, allowing employees to enter much of their own payroll information, and adopting some form of electronic payment, the payroll department will find that its role changes from data entry to monitoring, to ensure that incorrect information is excluded from the system. A flowchart of the revised payroll process is shown in Exhibit 7.4.

Some of the controls related to payroll are altered or rendered ineffective by these changes. The next section address alternative control systems that allow the accounting department to maintain adequate control over the more streamlined payroll process.

CONTROL ISSUES

Payroll is among the most heavily controlled of all accounting functions, since there is a risk of incorrect payment, and of outright fraud. The more streamlined processes advocated here introduce the prospect of new types of errors or fraud. For example, if a company were to issue bar coded badges to employees for a timekeeping system, employees could make a photocopy of the bar code on the back of the badge, incorporate it into a fake badge, and have someone else "buddy

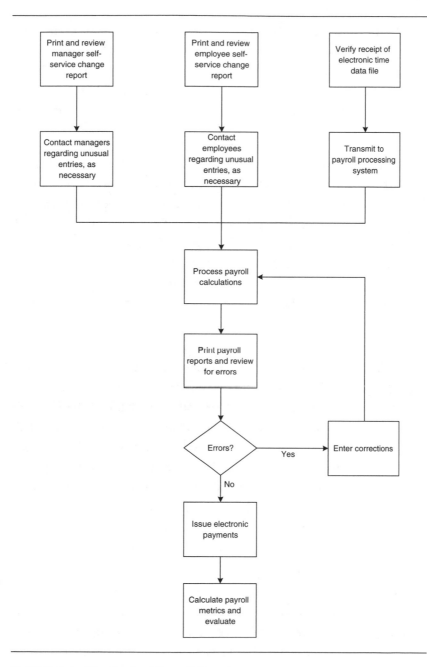

Exhibit 7.4 The Revised Payroll Process

punch" the fake badge into the scanner at any hour, resulting in more hours "worked." As another example, using direct deposit for payments makes it more difficult to spot payments to nonexistent "ghost" employees. This section discusses several control problems that arise as a result of the streamlining methods noted earlier, along with possible solutions.

- *Employee self service.* Having payroll clerks enter payroll changes into the computer system from signed deduction request forms keeps deductions from being illegally altered by people other than the employees requesting the change. If the control were removed, anyone could enter the deduction system from a computer terminal and alter key information. To mitigate this risk, employees should enter security codes to access the system. More protection can be added by having the computer system periodically require new passwords.

- *Employees can copy badges.* Having employees use a manual time clock to punch their time on a cardboard time card is a good control, because it creates a permanent visual record. When this is replaced by a bar-coded employee badge, the bar code can be copied and given to someone else, who then buddy punches on behalf of the employee. To prevent bar code copying, cover the bar code portion of the badge with a red laminate, which prevents a photocopier from detecting the bar code under the laminate. Also, mount a video camera next to each electronic time clock, so that employees will think they are being visually monitored when they use the clock. Even if the camera is not really being used, it may keep employees from attempting buddy punching.

- *Employees cannot see time card results.* When cardboard time cards are used, employees normally examine each time punch on the cards, to ensure that the time stamp occurred. However, with an employee badge, there is no such record, so the employee has no way of cross checking the entry. In an extreme case, a nervous employee may run his card through the machine several times in a row, effectively clocking himself in and out several times over. To avoid this problem, newer scanning systems have a lighted display that shows the name of the individual who just scanned a badge, as well as the time and date of the scan. Do not acquire lesser equipment that does not contain this feature.

- *The check signer is eliminated.* The check signer occasionally finds a mistake in payment amounts when signing paychecks. If direct deposit is used, then there is no check signer, so there may be a slightly increased incidence of errors. To avoid the issue, print the check edit report and review it before direct deposit information is sent to the bank; this report summarizes the entire payroll, and so is a good final edit check.

- *Employees do not review electronic remittances.* When employees receive their paychecks, they normally review the accompanying remittance advice to ensure that their pay was correctly calculated. However, if the remittances are stored electronically on a Web site that employees must access, it is extremely likely that the majority of them will not review the remittances. In some cases, employees may not review remittances more than once a year (and then only to download W-2 information for their tax returns). Thus, this control is compromised, and should no longer be considered effective.

- *Direct deposit does not allow visual receipt.* When employee pay is restricted to checks that are physically handed to employees, it is easy to determine if checks have been printed for employees who do not exist ("ghost employees"), since they are not on hand to receive payment. When direct deposit is used, this control disappears, because the payment is sent directly to a bank account instead. This makes it easier for payments to be made to ghost employees. To prevent this problem, one option is to still physically hand out the remittance advice; this means that the company will still lose one direct deposit that has just been made to a ghost employee, but will keep from making a second payment. However, continuing to hand out any paper-based material is a non-value-added activity. Also, it is possible to instead strengthen other controls, such as a continuing audit of employee existence, supervisory reviews of hourly time records, and the prompt entry of information about employees leaving the company.

In summary, when new streamlining methods are implemented, it is possible that the overall control structure will be weakened, especially surrounding the ability of someone to pay a ghost employee. Alternative controls will be needed to mitigate the revised risk structure.

COST/BENEFIT ANALYSIS

This section demonstrates how to conduct cost/benefit analyses for employee self-service, bar-coded time card systems, direct deposits, and electronic payroll remittances. The expected revenues and expenses used in these examples will vary considerably from a company's actual situation, but the format used is a good framework for a realistic cost/benefit analysis. Examples are as follows.

GIVE EMPLOYEES DIRECT ACCESS TO DEDUCTION RECORDS

The controller of the Circular File Company, manufacturer of wastebaskets, is concerned about the cost of having two employees who do nothing but update the deduction records of the firm's 1,375 employees. These two employees earn an average of $34,300 each. The controller decides to have the employees enter their own deduction changes. To do this, the company must subscribe to a new module of its third-party payroll processing system, which it can lease for an annual cost of $20,000. This module allows computer access by all employees to directly change their deduction records. Unfortunately, 90% of the staff comprises production workers with no access to a computer. Accordingly, the controller decides to install four computer kiosks at various locations in the facility for access by the production staff. These kiosks will allow computer access by all staff, and will cost $5,000 each to acquire and install. General training sessions will be held for the staff during their lunch breaks. The cost of the trainers and training materials will be $3,500. Should this project be initiated?

One-time project costs are for the computer kiosks and terminals, while there will be a permanent fee for access to the new software module. The savings are entirely from eliminated data entry payroll costs. The analysis follows:

Cost of Installing Deduction Interface	
Cost per computer kiosk	$5,000
Number of kiosks	×4
Total cost of kiosks	$20,000
Annual software fee	$20,000
Training cost	+ 3,500
	$23,500
Total cost of deduction interface	$43,500
Benefit of Installing Deduction Interface	
Cost per year of data entry clerk	$34,300
Clerical positions eliminated	×2
Total savings from deduction interface	$68,600

With costs of $43,500 and annual savings of $68,600, this project has a rapid payback, and should be implemented. A modest variable in this analysis is the training cost, which may be needed on a repetitive basis, and so will increase the cost.

INSTALL BAR-CODED TIME CARD SYSTEM

Katherine Peterson, general manager of Sucker Candy Corp., notices that the company employs an accounting clerk to do nothing but enter time card information for the firm's 347 production workers, who operate the firm's candy manufacturing equipment during all three shifts. This clerk also corrects mistakes made during the original data entry. Investigation with the various manufacturing supervisors reveals that one time card in eight is incorrectly entered by the clerk, resulting in complaints by the manufacturing staff. The controller contacts a manufacturer of bar-coded time card scanning equipment and finds that making a bar-coded, laminated time card will cost $1.50 for each employee. The firm will also need a pair of bar-code scanners, to be located next to each exit door. The scanners cost $1,800 each. The scanners will be linked to a dedicated personal computer, which collects and stores the information. The PC, including network hookup charges, costs $3,000. Cabling must also be purchased and installed to link the two scanners to the PC. The total cabling costs are $2,800. Finally, the controller must purchase interface software for $2,000 that links the payroll information on the PC to the payroll software. If this system is installed, the time card entry clerk will be reduced to half-time and will be assigned the task of reviewing time card entries on the time card PC for errors. The time card entry clerk earns $30,000 per year. Should the bar-coded time card system be installed?

The cost of the system is related to purchasing the time cards, scanners, cabling, PC, and software interface. These are all one-time costs, except for purchasing additional bar-coded time cards for new employees and replacing lost ones. There are ongoing savings from reduced data entry labor. The analysis follows:

Cost of Installing Time Card System

Number of production employees	347
Cost per time card	× $1.50
Total cost of time cards	$521
Cost per bar-code scanner	$1,800
Number of scanners needed	× 2
Total cost of scanners	$3,600
Cost of personal computer	$3,000
Cabling costs	+ $2,800

Cost of software interface	+ $2,000
Total time card-related costs	$11,921
Benefit of Installing the Time Card System	
Cost per year of data entry clerk	$30,000
Percentage of year no longer needed	× 0.50
Total labor savings	$15,000

With costs of $11,921 and annual savings of $15,000, the bar-coded time card system will be paid back within one year, and certainly should be accepted.

DIRECT DEPOSIT PAYCHECKS

Mr. Baumgartner, president of Sales on Call, a telephone wholesaler, receives many complaints from his sales staff, which is constantly traveling and therefore unable to pick up paychecks at the office on payday. Mailing the checks is not a viable alternative, for many salespeople are out of town so much that they cannot check their mail on a timely basis. This results in salespeople being unable to pay their credit card bills on time, which in turn creates a continuing credit problem while traveling on company business. To solve the problem, Mr. Baumgartner is considering using direct deposit to electronically transfer money directly into their bank accounts. The company employs 42 salespeople. The cost to make a direct deposit is $0.75 per paycheck deposited. The company pays employees twice a month. In addition, Mr. Baumgartner would like to wire payments to salespeople for their weekly expense reports. The entire sales staff travels every week, excluding their annual two-week vacations. If the direct deposit system is implemented, Mr. Baumgartner expects to stop issuing an average of $1,000 in advances to the sales staff to cover their expenses, because they have not received their paychecks. On average, 25 salespeople are in receipt of these advances at all times. The company borrows money at a 9.5% interest rate. Should Sales on Call implement a direct deposit system?

The cost of direct deposit is the transaction fee charged per payment. The savings are from the interest expense saved on advances made to salespeople. The analysis follows:

Cost of Direct Deposits	
Number of salespeople	42
Number of pay periods per year	$\times 26$
Number of pay deposits	1,092
Number of salespeople	42
Number of expense reports per year	$\times 50$
Number of expense deposits	2,100
Total number of deposits	3,192
Cost per direct deposit	$\times \$0.75$
Total cost for direct deposits	$2,394
Benefit of Direct Deposits	
Number of salespeople with advances	25
Outstanding advance per year	$\times 1,000$
Advances outstanding	$25,000
Interest cost	$\times 95\%$
Interest cost saved	$2,375

In short, with costs of $2,394 and savings of $2,375, there is no clear savings from implementing a direct deposit system. This is a common issue with direct deposit, especially when it is implemented for all corporate staff rather than just the salespeople who need it the most. Alternatively, this can be presented to management as more of an employee benefit than a direct cost savings to the company.

ISSUE ONLY ELECTRONIC REMITTANCE ADVICES

The controller of the EverKlear Windshield Company notices that the cost of mailing direct deposit remittances to all company employees is one dollar per paycheck, including postage costs. The company has 650 employees and a one-week pay cycle. The controller conducts a survey and finds that 75% of the company's employees have Internet access. By posting remittance advices on a secure Web site, the company can avoid the mailing costs associated with sending the information on a paper form. EverKlear outsources its payroll processing, and the payroll processor offers this added service for $0.45 per remittance advice. Should this project be completed?

The cost of the project is a largely variable charge, while the savings are from the labor, envelopes and paper, and postage needed to mail out remittances. The analysis follows:

Cost of Issuing Electronic Remittance Advices	
Number of employees	650
Percentage of employees with Internet access	× 75%
Number of employees with Internet access	488
Electronic remittance fee per paycheck	× $0.45
Total cost per pay cycle	$220
Number of pay cycles per year	× 52
Total cost of remittance mailings	$11,440
Benefit of Issuing Electronic Remittance Advices	
Number of employees	650
Percentage of employees with Internet access	× 75%
Number of employees with Internet access	488
Mailing cost per employee per remittance advice	× 1.00
Total cost per pay cycle	$488
Number of pay cycles per year	× 52
Total cost of remittance mailings	$25,376

Since all costs and benefits associated with this project are variable, it is evident that the project will be profitable as of the very first payroll, and so it should be implemented at once.

REPORTS

The payroll department is usually awash in reports, so it is not a service to that department to come up with even more reports. However, there are a few simple ones that contribute to the efficiency of the department.

The first report is a *list of employees not using direct deposit*, or for whom direct deposit pre-notes have failed. Since using direct deposit reduces the workload of the department, the payroll manager should print this list of non-users regularly and attempt to persuade them to accept such payments.

If the payroll department has achieved the transfer of all data entry to the company's employees and managers, then it now shifts into a transaction monitoring role, where it uses a variety of special reports to comb through the data for incorrect entries, only showing transactions that fall outside the normal pay boundaries. Any of the following reports are useful for this purpose:

- *Active employees with no payments*. Indicates that timekeeping records may be missing.

- *Time entered for inactive employees*. Indicates that either an inactive employee has come back to work (which requires a variety of deduction activations), that an employee has charged time to the wrong employee code, or that some other employee is attempting to falsely record time on behalf of the inactive employee.

- *Leave balances are negative*. Indicates that an employee is taking more leave than is authorized. This calls for supervisory approval.

- *Negative deductions*. Indicates when a payment is being made to an employee via a negative deduction, requiring review for underlying causes.

- *Negative taxes*. Indicates either negative pay situations or cases where previous excess tax deductions are being corrected.

- *Hourly rate less than minimum wage*. Indicates that a payment is being made that is below the legal pay limit. This report may have to be sorted by state, because some state-mandated minimum wage rates are higher than others.

- *Hourly rater greater than $____.* This report is designed to catch a very large hourly rate exceeding what any employee normally would receive.

- *No pay greater than $____.* This report is designed to catch very large payments exceeding what any employee normally would receive.

In short, a payroll department that primarily monitors the date entry of other people is primarily looking for exceptions, and so can use a variety of specialty exception reports to comb through the data for possible errors or instances of potential fraud.

METRICS

The following measurements are useful for tracking the cost and payment speed of the payroll system.

TOTAL PAYROLL DEPARTMENT COST

If the amount of data entry and pay distribution effort by the payroll staff declines, then the department's overall costs should also drop. Consequently, one should track the department's total cost per employee on a trend line. The calculation should be the total direct cost of the department (i.e., with no corporate overhead charge), divided by the total number of employees in the company. This is a better measure than a simple analysis of total department costs, since costs should vary in approximate proportion to the number of total company employees.

PAYROLL DEPARTMENT CHANGES

A key goal for the payroll manager is to eliminate all data entry by the payroll department. Thus, a good measurement is to track the total

Exhibit 7.5 Error Totals by Payroll Transaction Type

Transaction Type	Total Number	Number of Errors
Address entry	9	1
Deductions entry	8	0
Employee addition/deletion	42	7
Exemptions entry	15	0
Hours entry	312	49
Status entry	3	0
Totals	389	57

number of entries made by the payroll department, and to also categorize them by type. The manager can then work on reducing the various types of changes.

PAYROLL TRANSACTION ERROR RATE

If there is a transaction error, it usually requires an inordinate amount of time to find and correct it. Thus, it is useful to measure the total transaction error rate both on a trend line and by transaction type, to spot spikes in errors and to research which error types are causing the trouble. To calculate the payroll transaction error rate, divide the total number of payroll errors for the payroll cycle by the total number of payroll transactions made during that cycle. The table in Exhibit 7.5 shows the best format, since it reveals error rates for specific types of transactions.

PROPORTION OF PAYROLL ENTRIES TO HEADCOUNT

The payroll department may be tracking too much information about employees through the payroll system, which requires an excessive number of payroll entries to maintain, leading to reduced efficiency. By compiling the total number of payroll entries in the system per employee, one can see if this "data load" is excessive. The formula is to accumulate the number of all payroll deductions, memo entries, and goal entries and divide this total by the number of full-time equivalents. The formula is:

$$\frac{\text{Total deductions} + \text{Total memo entries} + \text{Total goal entries}}{\text{Total number of full-time equivalents}}$$

In summary, a few simple measurements are needed to provide an effective set of statistics about the performance of the payroll department. Collecting information for these measurements is not difficult, and provides management with enough information to spot department efficiency problems quickly.

SUMMARY

This chapter concentrated on improving the efficiency of the data entry portion of payroll, since there are many detailed manual calculations and keypunching required at that stage. In particular, the implementation of employee self service, manager self service, and computerized time clocks can eliminate the bulk of the payroll department's work in this area. A variety of improvements later in the payroll process can make payroll the most automated of all accounting functions. This fundamentally changes the role of the payroll department, from one whose primary task is data entry, to one that primarily monitors entries made by the rest of the company.

Chapter 8

The Budget

The process of creating the annual budget consumes a large amount of managerial time, not only to determine budget assumptions but also to decide on sales projections, the need for additional facilities, financing needs, staffing concerns, and to critique the plans of other departments. This series of meetings may be repeated at various times during the year if revenues depart so far from the budget that the associated expense budgets are no longer relevant. Further, other meetings may take place that focus on how actual revenues and expenses are turning out in comparison to the budget. In total, the time allocated to the budget process is among the largest draws upon the time of today's manager. This chapter discusses how to reduce the time needed to create, modify, and monitor the budget.

CURRENT SYSTEM

This section gives an overview of the budget process. It shows how the various types of budgets are linked, and how a change in one budget causes changes in other budgets. Also, it notes the inefficiencies that are built into the typical budget process.

The budget cycle begins with a management meeting that reviews the firm's strategic objectives. These objectives may involve such issues as eliminating entire businesses or creating new ones, and so have a major ripple effect throughout the budget. The strategic objectives meeting then leads into a (usually prolonged) discussion of revenue objectives for the upcoming year. This is a lynchpin of the budget, since many expenses are based on revenue. The revenue goal also requires a detailed analysis of sales targets for sales regions and salespeople, which may entail a change to the commission structure and a change in the number of salespeople, the method of selling, and an expansion into entirely new sales territories.

With a preliminary sales plan completed, the production staff now works on a production plan. This plan determines how many materials must be purchased, how many workers are needed, and whether additional equipment or facilities must be constructed to increase capacity to meet the number of units required by the sales plan. If management deems the resulting cost of goods sold to be excessive, then the industrial engineering staff may be brought in to plan for revised production systems.

Once there is general agreement on the production plan, it is handed to the other company departments, such as engineering, administration, and information technology (IT). These departments construct budgets based on the revenue and production plans. These plans include expenses (typically with a heavy emphasis on personnel costs) as well as fixed asset purchases. When combined with facilities requirements that are included in the production plan, the budget team has enough information to construct a capital expenditures budget. This concludes the primary set of budgets, with one exception.

The final budget is the financing plan. This budget takes the cash flow information from all previous budgets and the cash requirements of the capital expenditures plan, and derives working capital needs based on days of receivables and payables, as well as inventory turnover. With these inputs, the financing plan derives either the amount of cash spun off by the company or the amount required. This information is reported back to the management team, which frequently alters the budget to take into account the realities of financing. A graphical representation of the budgeting process is shown in Exhibit 8.1.

In summary, the typical budget process requires a large number of iterations and many meetings by the management team before a satisfactory budget is created. This is not an efficient use of managers'

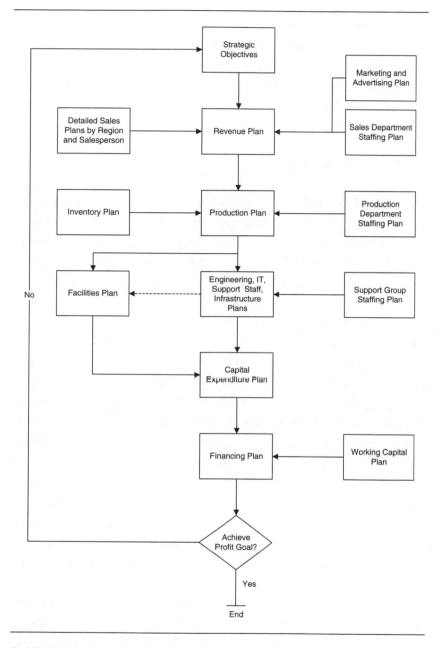

Exhibit 8.1 Typical Budget Process

time, since they must meet many times to make incremental changes to the budget, and then wait for the budget team to process the changes and return the results to them. The following section discusses ways to reduce the number of budgeting iterations, as well as ways to reduce the workload of everyone involved in the budgeting process.

REVISED SYSTEM

This section focuses on streamlining the budget process. This can be done by altering the format of the general budget model, creating a detailed budget structure that greatly improves budgeting efficiency, and then conducting high-speed budget iterations. When implemented, these changes will allow a company to prepare a budget without an excessive commitment of management time.

The first order of business is to determine the type of budget model to be used, and around which all other budgeting improvements will be made. The primary decision is whether a company will use a homegrown budgeting spreadsheet, or *purchase a budgeting software package*. In multi-location companies, a common budgeting technique is to send a budget template to each location, and ask the local manager to complete the budget for that subsidiary. The trouble is that the local managers may add or subtract accounts, which makes it much more time-consuming for the headquarters budget analyst to shoehorn the resulting budget back into the master budget. To avoid this, buy a commercially-available budgeting software package that users must access for data entry purposes. This allows the company to lock down accounts, and also requires subsidiaries to enter budget information directly into the system, thereby avoiding data entry labor by the budget analyst. However, the budget analyst's time now shifts to monitoring the accuracy of the entered information, into which errors may have been introduced. Also, budgeting software is expensive.

If a company elects not to buy budgeting software, it will construct its budget using an electronic spreadsheet. This is one of the most complex spreadsheets that a company maintains, since it includes a multitude of variables and cross-linked formulas. It is extremely easy to introduce an error into the budget model, which may substantially throw off its results, and which also requires a great deal of time to track down. Here are some techniques for reducing the number of errors in a spreadsheet model:

- *Keep a copy.* Archive a fully-tested copy of the budget, which can be used in case subsequent budget versions prove to be excessively flawed.

- *Reduce the number of accounts.* If there are fewer account line items, there are fewer cells into which errors can be introduced.

- *Summarize small-dollar departments.* If a department or cost center has a negligible dollar total, then allow it just a single expense line item.

- *Change cell color.* A budget contains some cells that are calculations, and some that are manual entries. The difference between these two types of cells may not be clear to someone who is rapidly entering modifications. To avoid having a calculation cell erased, alter the color of the text in each cell. For example, a data entry field can be blue, and a calculation field can be red.

- *Lock down spreadsheets.* It is also possible to lock down entire spreadsheets, thereby keeping unauthorized people from altering the spreadsheet.

- *Use formulas for variable expenses.* If a key variable in a budget model (such as revenue or headcount) is altered, this requires a multitude of related changes to all of the expenses that vary with the key variable. It is very difficult to track down and manually correct each of these expenses. Instead, set each one up to be a fixed percentage of the related key variable, so that they change automatically.

- *Cluster key variables in one place.* A poorly-designed spreadsheet has key variables spread throughout the spreadsheet, where they can easily be missed. For example, the amount of wages applicable to Social Security taxes may be buried on a payroll worksheet, while the consumer price index is listed on a departmental expense worksheet. Instead, put them all in one place where a budget analyst can more easily monitor and alter them.

Having decided upon the general type of budget model, the next question is whether to initially operate with a summary-level or detail-level budget.

A common early step in the budget process is to have all department managers submit detailed budgets. However, this is putting the cart before the horse, for the general activity level (i.e., revenues) must be determined before anyone should waste time with detailed budget planning. Thus, *do not budget detailed expenses until a summary-level budget is approved.* Until the summary-level budget is approved, department managers can submit summary-level budgets in response to various revenue-level iterations. These summary budgets can be roughly compiled merely by tracking the number of employees in the department; the largest cost of most departments relates to salaries and benefits, with other costs such as office supplies, varying in approximate proportion to the number of people in the department. Therefore, re-budgeting based on department headcount is fast and yields reasonably accurate results, without conducting a line-by-line review of all budgeted expenses. A departmental budget based on headcount is shown in Exhibit 8.2.

The high-level budget should include an analysis of working capital. By doing so, management understands the incremental cash requirements associated with strong forecasted sales growth; this frequently results in a downward recasting of the budget to find an activity level that is supportable with less additional cash. Working capital can be determined even without a complete balance sheet. A simple listing of receivables, payables, and inventory is sufficient for a quick working capital analysis, such as the one shown in Exhibit 8.3.

Since the obvious path thus far is to work with a summary-level budget, it also makes to *reduce the number of accounts.* This can be pursued throughout the year rather than just during the budget process. The intent of having fewer accounts is that there are fewer line items

Exhibit 8.2 Engineering Department Budget Based on Headcount

Job Title	No. of Employees	Average Salary	Extended Salary	Benefits Percentage	Total Employee Cost
VP Engineering	1	$125,000	$125,000	20%	$150,000
Engineering Super.	3	75,000	225,000	20	270,000
Engineer III	15	60,000	900,000	20	1,080,000
Engineer I	15	45,000	675,000	20	810,000
Designer	12	35,000	420,000	20	504,000
Librarian	2	30,000	60,000	20	72,000
Secretary	3	25,000	75,000	20	90,000
Total staff	51		$2,480,000		$2,976,000

Variable Expense Items	Cost per Employee	No. of Employees	Total Variable Cost
Office supplies	$450	51	$22,950
Travel	2,570	51	131,070
Telephones	350	51	17,850
Total variable			$171,870

Fixed Expense Items	Total Fixed Cost
Subscriptions	$500
Dues	1,000
Annual awards	350
Total fixed	$1,850

Exhibit 8.3 High-Level Working Capital Analysis

Account Type	This Year Turnover	Next Year Turnover	This Year Working Capital	Next Year Working Capital
Revenue			$30,000,000	$50,000,000
Cost of goods sold			9,000,000	15,000,000
Receivables	45 days	40 days	3,750,000	5,555,000
Inventory	10 turns	12 turns	900,000	1,250,000
Payables	35 days	35 days	875,000	1,458,000
Total working capital			$3,775,000	$5,347,000

in the budget that must be estimated, fewer line items that can be incorrectly entered, and fewer line items that can be incorrectly summarized by bugs in the budget model. In short, fewer accounts shrink the work of preparing the budget.

A further improvement in the efficiency of the budget model is to *use flexible budgeting* for as many accounts as possible. A fixed budget derives a profit based on a very specific revenue level for the year. Expenses are tied to that revenue level, and must be manually changed every time another budget iteration results in a new revenue level. With flexible budgeting, it is possible to designate both fixed and variable portions of some expenses and have the variable portions change as revenue changes. This allows the company to quickly revise the budget with minimal manual effort. It must be noted, however, that flexible budgeting only works within a relatively narrow range of revenue figures—once revenue departs significantly from the expected target, there may be a significant change in the related expenses in terms of headcount and facilities that will require a complete recasting of the budget.

By implementing the preceding recommendations, we now have a general budget model that can be efficiently updated through many iterations. The succeeding changes will allow us to create planning systems and a more detailed budget structure that will further improve the efficiency and effectiveness of the overall process.

The most crucial of all budget meetings is the one that occurs the least frequently—*the strategy and assumptions meeting*. This meeting brings together experienced representatives from all major departments to critique the major assumptions of the budget for the year before any detailed work is performed at all. This meeting keeps the budget staff from spending too much time creating detailed plans based on false assumptions. By ensuring that the correct high-level direction is being taken right from the start, less time will be wasted on constructing the budget.

Another important strategic issue is that a company should *immediately limit the budget with the maximum available funding*. This is necessary to keep management from spending considerable time devising a strategy that will require immense capital resources that cannot realistically be obtained. By setting a funding cap at the start of the process (usually in the form of a debt/equity ratio or hard cap on the number of new shares to be issued), a company can avoid a first budget iteration that is clearly unobtainable.

Most budgets contain a significant number of projects. If funding in the budget is not sufficient to allow all projects to be completed, there is usually a great deal of time-consuming haggling among managers to determine which projects are to be funded and which are to be delayed. To avoid this problem, one of the first tasks in the budget process is to *determine project degree of fit with strategic goals*. If there is no close strategy link, then a project should be abandoned. By making projects pass through this analysis phase early in the budget process, one can promptly eliminate several from consideration, without any additional need for detailed project justifications.

When creating a budget that contains major revenue changes from existing levels, a common problem is failing to reflect this change throughout the budget, resulting in an inadequate budget for additional staffing, machinery, and facilities to handle the added growth. For example, a planned increase in revenue requires a corresponding increase in the number of sales staff who are responsible for bringing in the sales, not to mention a time lag before the new sales personnel can be reasonably expected to acquire new sales. As another example, if the cost of direct labor has been budgeted to decline as a result of increased automation, has investment in automation been sufficiently budgeted, and has a suitable time lag been built into the plan to account for the ramp-up time needed to install the new equipment? To avoid these issues, *define capacity levels in the budget model*. This can take the form of a table within the budget that notes the capacity level of various positions. For example, there must be one computer help desk person for every 250 employees, one salesperson for every $1 million of sales, and one machine operator per work cell. It is very important to list these capacity levels for previous years in the same table, thereby providing a frame of reference that tells the reader if the assumed capacity levels in this year's budget are attainable.

The issue of capacity planning also brings up the need for proper planning for step costs. Step costs are blocks of significant additional expense which must be incurred when a certain level of activity is reached. For example, machinery can only operate at a reasonable capacity level, perhaps 75%, before another machine must be added to cope with more work, even if that workload will only fill the new machine at a very low level of capacity. The same principle applies to adding personnel or building space. In all cases, there is a considerable added expense that must be incurred in one large block. If the expense is sufficiently large, it can play havoc with the total level of expenses. The best way to determine when an increase in step costs will occur is to *create a table of activity measures for key step costs*. For example, a new piece of production equipment is needed when throughput reaches 10,000 units per month. The budget model should then indicate the need for an additional capital investment once the planned production volume reaches 10,001 units per month.

Another item that should be derived early in the budget process is *the amount of fixed costs*. A fixed cost is any cost that is likely to exist during the upcoming year unless management takes specific action to remove or add to the cost. Examples of these costs are utilities, rent or lease payments, depreciation on existing assets for the upcoming year, taxes, and employee benefit costs. This information can be prepared far in advance of the rest of the budget information to save time. By doing so, management can review the information early in the budget cycle to take prompt action if adjustments are needed.

Determining *cost drivers* allows the company to develop a budget that reacts well to changes in activity levels. A cost driver is an activity that changes a cost. If a company finds, by constantly comparing budgeted costs to actual costs, that specific activities dramatically change costs, then those drivers should be budgeted. For example, the number of employees onsite may have a direct effect on utility costs. Therefore, the budget should link the number of onsite employees to changes in the utilities expense. Similarly, the number of employees can be linked to the cost of office supplies. This cost driver yields a more accurate cost than linking either utilities or office supplies to changes in revenue. By using cost drivers, the budget becomes more refined and requires fewer laborious changes and variance analyses over its life.

If a company has an ongoing process improvement plan, it may have guaranteed its employees that there will be no layoffs resulting from the plan. Usually, this means that employees freed up from their normal tasks are shifted into permanent process improvement mode, possibly as part of a company-wide team. If so, they are no longer working for their original department, and should be shifted into a new cost center. Otherwise, retaining employees in their original departments does not delineate the amount of labor-related savings already generated, and may even undermine the improvement effort, since there is little financial incentive by department managers to look for additional improvements.

A rarely-used efficiency improvement is *reducing the budget from a monthly to a quarterly plan*. Top-level management budget meetings can frequently degenerate into an argument about the specific months in which revenues and related expenses will occur. If the number of budget periods are cut to one-quarter of the previous number, the amount of discussion related to the exact placement of revenue and expenses can also be reduced. For those companies needing budget vs. actual reports on a monthly basis, consider formulating the budget on a quarterly basis, and then dividing each quarter by three in order to obtain exactly the same budget for the three months within each quarter.

Up to this point, we have concentrated on building systems that will improve the basic process of constructing a budget. But what about future budget iterations? What can be done to condense and reduce any successive budget formulations?

An excellent way to introduce some structure to budget iterations is a *budget procedure*. This procedure, distributed at the initiation of the budget cycle, includes due dates for all budget deliverables and who is responsible for each one. A sample budget should also be included, as well as sample budget forms and a calendar listing all due dates. This degree of organization is particularly necessary for companies with subsidiaries, where there are many people in outlying areas who must send in their budget information before the overall budget can be created. Even in a smaller company where the budget analyst is within reach of all departments, it is still a good idea to introduce some rigor to the process, so that there can be no excuses by the staff regarding the lateness of budget information.

The traditional way to construct a budget is to send the budget procedure and a budget template to each manager who is expected to provide input to the model. The expectation is that the manager will complete a "from scratch" budget and return it to the budget analyst. However, filling out budget forms requires a considerable amount of management time, and is a prime reason why the budgeting cycle is roundly detested by many managers. A more efficient approach is to have the budget analyst *preload budget line items*. Most expenses are relatively fixed from year to year, and are easily linked to key drivers, such as head-count. Consequently, the analyst can probably arrive at more accurate budget numbers than a department manager for most line items. This approach leaves only a few of the larger and more variable accounts for managers to derive. In some cases where a department is expecting no major changes for the next budget year, it may even be possible for the budget analyst to create an entire department budget, which a manager only needs to review.

Once an acceptable budget has been agreed upon, the budget team has one remaining task. It must *ensure that performance measurement and reward systems are in place*. This is necessary to ensure that the company's performance against the budget is tracked and fed back to management, and that certain employees are rewarded for their ability to meet budget objectives. Without a performance tracking and reward system, the company will be more likely to <u>not</u> achieve budget goals, necessitating a recasting of the budget later in the year to bring expenses into line with a reduced level of corporate activity.

In summary, the most time-consuming part of the budget process is the number of iterations required to arrive at a budget model that is workable in light of funding needs and corporate resources. This section addressed a number of ways to create an easily maintainable and efficient budget model, thereby eliminating a large part of the expense and wasted time associated with creating several budget versions. By doing it right the first time, the cost of creating a budget can be significantly reduced. A streamlined budget process is shown in Exhibit 8.4.

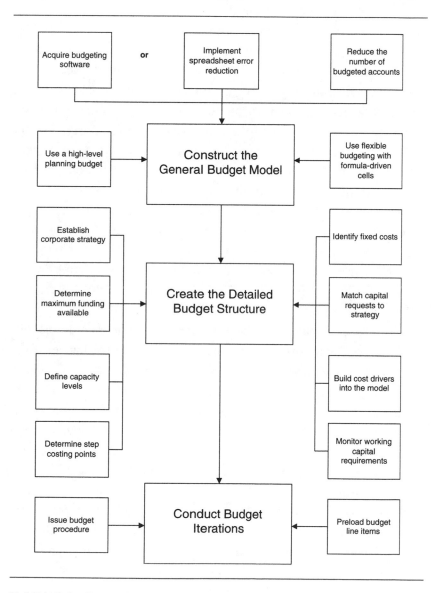

Exhibit 8.4   Streamlined Budget Process

CONTROL ISSUES

Unlike many of the other accounting-related areas covered in this book, the budget cycle has no transactions. Since there are no transactions, there is no need for controls in the budget area.

However, it is important once the budget has been finalized that expenditures in excess of the budgeted amount are highlighted for the benefit of management. There are several ways to accomplish this. First, the company's general ledger software should include a feature that lists the budget amount for each account next to the actual expense for the month and the year-to-date. Always use the budget comparison when reporting financial results to managers. The second way to track expenditures in excess of the budget is to have a drill-down feature in the accounting software. A drill-down feature presents the user with the detailed costs that make up a summary cost balance, usually by jumping directly from the summary screen to a detailed expenditure screen. A third technique is to send a report to management that lists all expenditures over a certain dollar amount during the period, with a detailed description of the expenditure.

Also, once the budget has been finalized, it is important to institute controls over major expenditures that were noted in the budget. Even though the budget has already been approved by management, individual large purchases or pay increases should still be justified to management in detail and not implemented without the approval of management. This is because the intent of the type of budgeting presented in this chapter is to create a budget without getting into extreme detail, thereby allowing the management team to complete a quality budget very quickly. If capital expenditures that were listed in the budget at a macro level (e.g., "unspecified production equipment for $100,000") are implemented without further controls, then this defeats the purpose of a quick budget, which is to decide on an appropriate level of corporate activity for the upcoming year, rather than to act as an expenditure approval document.

In short, it is important to track expenditures as they compare to budgeted expenses and to institute tight management review controls over spending amounts that were approved *in concept* during the budget formulation process.

COST/BENEFIT ANALYSIS

This section contains two cost/benefit analyses for streamlining the budget model and reducing the budget review time. The expected revenues and expenses used in these examples will vary considerably from a company's actual situation, but the format used is a good framework for a realistic cost/benefit analysis. Examples are as follows:

SIMPLIFY THE BUDGET MODEL

The controller at fast-growing InterMode, a maker of model trains, is concerned that the budget process is taking too much of the financial analyst's time each year. By altering the budget model to include fewer accounts, varying some expenses directly as revenue fluctuates, and cutting back reporting periods to quarterly, it should be possible to reduce the hours that the analyst devotes to the budget. Further examination reveals that, in the previous year, the budget went through four iterations that required reentry of information into the budget for all four iterations. The analyst estimates that it took two days to revise the budget on each occasion. By making the preceding changes, the analyst should be able to cut the reentry time in half. However, it will require two days of work to alter and test the budget model to ensure that the simplifications do not cause any errors. The financial analyst earns $65,000 per year. Should InterMode alter its budget model? The analysis follows:

Cost of Streamlining the Budget Model

Number of days to alter budget model	2
Daily cost of analyst	$\times$ $250
Total cost of new budget model	$500

(*Continued*)

Benefit of Streamlining the Budget Model

Number of budget iterations/year	4
Number of days to alter budget model	× 2
Total days/year to alter budget model	8
Time savings from new budget model	× 50%
Number of days saved by new budget	4
Daily cost of analyst	× $250
Total savings from new budget model	$1,000

With costs of $500 and savings of $1,000, the cost of budget alteration achieves break even after only two budget iterations and saves
money thereafter. However, the costs and savings are not "hard,"
since the financial analyst is presumably on salary and will not be paid
less if the budget is completed in less time. However, it does allow the
analyst more time to work on other projects, or perhaps an additional
iteration of the budget. Also, some budget models are extremely complex and may require far more revision time than the interval noted in
this example. For example, switching to a flexible budget, with the
large number of attendant formula changes, is very time consuming.

REDUCE BUDGET REVIEW TIME

The managers of Gregorian, Inc., producers of business calendars, are
bothered by the amount of time they spend preparing the budget each
year. The CFO recommends a number of changes to the process, including setting available funding levels in advance, eliminating detailed expense reporting until the high-level budget is complete,
prioritizing projects based on how they relate to corporate goals, and
linking the budget to a performance measurement and reward system.
These changes are also free to implement, since the only cost is three
days of the CFO's time to construct a new budget flowchart and a reward system. With these changes, the CFO feels the management team
can cut the number of budget iterations from seven to three, though
the typical three meetings per iteration will probably increase in

duration from three to four hours. There are six members of management on the budget committee, and their pay averages $100,000. The CFO is one of the six members of the budget committee. Should these changes be implemented?

The number of hours saved by reducing budget iterations must be offset against the time of the CFO to create the budget flowchart and reward system. The analysis follows:

Cost of Implementing Budget Review Changes	
Days of CFO time needed	3
Cost/day of CFO	× $385
Total cost of implementing changes	$1,155
Benefit of Implementing Budget Review Changes	
Number of budget iterations eliminated	4
Meetings/iteration	× 3
Total number of meetings eliminated	12
Number of people attending each meeting	× 6
Total savings from people attending meetings	72
Time/meeting	× 3 hours
Total number of attendance hours saved	216 hours
Less: increased length of remaining meetings	−54 hours
Net number of meeting hours saved	162 hours
Cost per hour per manager	× $48
Total savings from reduced budget time review	$7,776

With costs of $1,155 and savings of $7,776, there is a strong argument in favor of implementing the budget streamlining project immediately.

REPORTS

No periodic reports are needed besides the budget. However, the budget process can be assisted by a brief budget procedure that outlines due dates and responsibilities. This section includes an example of a very brief budget procedure for a company with no subsidiaries (which would otherwise greatly expand the bulk of the model).

A budget procedure should include, at a minimum, itemized budget deliverables, their due dates, to whom they should be sent, and who is responsible for completing them (see Exhibit 8.5).

The following are some observations regarding the budget procedure.

* *Length*. The procedure does not need to be very long to lay out a reasonable listing of due dates and responsibilities.

* *Job titles*. The procedure does not list the names of individuals who are responsible for various deliverables, since they may switch to new jobs. Instead, to keep the accounting department from having to reissue the procedure every time a new person becomes involved in the process, the job title that is responsible for each deliverable is used.

Exhibit 8.5   Budget Procedure

Ajax Symphonic Recordings, Inc.
2009 Budget Procedure

Send Deliverables to Budget Analyst   **Page 1 of 1**

Step	Date	Responsibility	Deliverable
1	10/05	Budget committee	Review strategic direction
2	10/15	Sales Vice President	Present revenue plan
3	10/20	Marketing Vice President	Present marketing plan
4	10/23	Production Vice President	Present production plan
5	10/28	Engineering VP, IT VP, CFO	Present engineering, administration, and IT plans
6	11/02	Production Vice President	Present facilities plan
7	11/09	CFO	Present capital expenditures plan
8	11/11	CFO	Present financing plan
9	11/20	Budget committee	Budget reiteration meeting
10	11/25	Budget committee	Budget reiteration meeting
11	11/30	Budget committee	Final management review meeting

- *Procedure date.* List the date of the procedure so that users will be able to compare different issues of the procedure and know which is the most current.

- *Procedure page.* List the number of pages in the procedure, in case an employee does not complete a budget step by being unaware that a page is missing.

- *Task dates.* It is difficult to reissue a budget procedure year after year without making changes to it, because the due dates may not fall on work days every year. Thus, list dates as specific work days of the month (e.g., the second Tuesday of November).

The budgeting procedure may be ignored by the people involved in submitting information to the budget analyst, or they may not even be aware of it. To avoid this problem, consider an annual rollout of the procedure, such as a training videotape, personal training session, or memo.

The focus of a summary management budget model is the information summarized in its tables. For this example, Exhibits 8.6 through 8.15 display the essential information needed by management to determine changes to the budget. Since there are very few expense line items in this model, any results from the model will be approximate. Nonetheless, it can be used to test the viability of a new set of

Exhibit 8.6 Summary Budget: Revenue and Cost of Goods Sold

Revenue	
Annual	$16,000,000
Monthly	1,333,333
Cost of goods sold per revenue dollar	
Materials	$0.217
Freight	0.03
Cost of goods sold, per year	
Materials	$3,472,000
Freight	480,000
Total	$3,952,000

Exhibit 8.7   Summary Budget: Department Payroll Cost

Department	No. of Employees	Average Salary	Total Salary Cost	Benefits Percentage	Total Payroll Cost
Production	120	$39,000	$4,680,000	20%	$5,616,000
Sales	12	78,000	936,000	22	1,142,000
Marketing	2	65,000	130,000	22	159,000
Engineering	18	70,000	1,260,000	22	1,537,000
IT	3	75,000	225,000	22	275,000
Administration	7	39,000	273,000	22	333,000
Total	162		$7,504,000		$9,062,000

Exhibit 8.8   Budget Model: Department Costs That Vary by Headcount

Department	No. of Employees	Cost per Employee	Total Employee-Variable Cost
Production	120	$1,100	$132,000
Sales	12	15,800	189,600
Marketing	2	1,500	3,000
Engineering	18	1,250	22,500
IT	3	3,100	9,300
Administration	7	2,200	15,400
Total	162		$371,800

Exhibit 8.9   Budget Model: Department Costs That Vary by Revenue

Department	Cost per Revenue Dollar	Total Revenue-Variable Cost
Production	$0.030	$480,000
Sales	0.025	400,000
Marketing	0.005	80,000
Engineering	0.008	128,000
IT	0.010	160,000
Administration	0.007	112,000
Total		$1,360,000

Exhibit 8.10   Budget Model: Department Fixed Cost

Department	Total Fixed Cost
Production	$108,000
Sales	25,000
Marketing	8,000
Engineering	62,000
IT	80,000
Administration	42,000
Total	$325,000

Exhibit 8.11   Budget Model: Profit

Revenue	$16,000,000
Cost of goods sold	3,952,000
Department payroll cost	9,062,000
Department employee-variable cost	371,800
Department revenue-variable cost	1,360,000
Department fixed cost	325,000
Total cost	$15,070,800
Net pre-tax profit	929,200
Taxes (35%)	325,220
Net after-tax profit	$603,980

Exhibit 8.12   Budget Model: Working Capital

Accounts receivable (45 days)	$2,000,000
Inventory (12 turns)	329,300
Total	$2,329,300
Accounts payable (30 days)	171,400
Net working capital	$2,157,900

management assumptions in a few moments, which allows management to rapidly cycle through a number of high-level iterations. Fast iterations keep the budget analyst from having to recalculate the budget details an excessive number of times.

Exhibit 8.13   Budget Model: Capital Spending

1	Manufacturing equipment upgrade	$25,000
2	Quality monitoring system	80,000
3	Online receiving system	15,000
4	Bill of materials software	7,500
5	Clean room construction	20,000
6	New lathe	12,000
7	New glass cooling tank	18,500
	Total	$178,000

Exhibit 8.14   Budget Model: Funding Available

Working capital required	$2,157,900
Capital spending required	178,000
Total	$2,335,900
Cash flow from profit	603,980
Depreciation	210,000
Total	$813,980
Funds required	1,521,920
Funds available	1,750,000
Excess funds available	$228,080

Exhibit 8.15   Budget Model: Performance Measurements

Department	Goal	Measurement
Sales	Sales/salesperson	$1,333,333
Marketing	Variable expense/employee	1,500
Production	Variable expense/employee	1,100
Engineering	Variable expense/employee	1,250
IT	Variable expense/employee	3,100
Administration	Variable expense/employee	2,200

This model would require expansion if a company had many more departments or subsidiaries; the example assumes just a few departments and no subsidiaries. Also, more detail may be necessary, depending on the variety of products sold, especially if the gross margin on different products varies appreciably.

The following points describe key issues within the budget model:

- *Revenue (Exhibit 8.6).* Many of the expense amounts listed later in the budget are tied to the revenue figure, and will automatically change as the revenue level changes.

- *Cost of goods sold (Exhibit 8.6).* This line item only includes materials and freight costs, since these are the only costs that vary directly with revenue levels. Production labor tends to be relatively fixed, especially if the work force is highly skilled, and so only varies as a result of a specific hiring or layoff decision. Also, overhead costs are recorded within the various departments for the same reason.

- *Department costs (Exhibits 8.7–8.10).* Payroll costs are listed first, since they are usually the largest department costs. Following payroll are costs that vary with the number of employees. Next are listed costs that vary directly with changes in the revenue level, such as production supplies in the production department and commissions in the sales department. Finally, fixed costs, which do not vary at this revenue level or with the number of employees, are listed. The portion of the fixed costs total ($325,000) attributable to depreciation is $210,000.

- *Working capital (Exhibit 8.12).* The days of accounts receivable and payable as well as expected inventory turns are noted as variables, and the amount of working capital required is calculated based on the revenue and cost of goods sold. In calculating the inventory turnover rate, the cost of goods sold includes the cost of materials and freight. Working capital required, plus the capital spending amount, yields the total amount of funding required for the upcoming year.

- *Capital spending (Exhibit 8.13).* The funding required by all company projects for the upcoming year is listed in this table. The projects are prioritized in order of how well they support the company's strategic goals or throughput bottlenecks for the upcoming year. The capital spending total, together with working capital requirements and profits, yields the total funds required for the upcoming year.

- *Funding available (Exhibit 8.14).* The maximum amount of funds available is listed next to the amount of funds required. If the amount required exceeds the available amount, management will probably want to recast the budget.

- *Performance measurements (Exhibit 8.15).* A key set of performance measurements can vary with the size of the budget to inform management of the goals that must be achieved to ensure that the budget model will succeed. These performance measurement calculations should vary automatically as the levels of revenue and expense change. The controller should have input into the metrics used, since he is the one who must track them.

In summary, a brief budget procedure is useful for informing employees involved in the budget effort of the due dates and deliverables required of them. This procedure keeps key information from being delivered late. Also, the summary management budget model, though abridged, is useful for determining the balance of costs and revenues needed by the company to achieve its profit goals within specific funding constraints. The summary model allows a quick review of revenue and expense activity for the upcoming year, which reduces the time managers must spend in the budget process. It also shrinks the time required by the budget analyst to complete the budget at a detailed level.

METRICS

There are a small number of metrics available for measuring the efficiency of the budget process. First, track the time period covered by the budget process by recording the start and finish dates. It is simplistic, but can be compared to the budget duration in previous years to see if the company has experienced any success in compressing its budget cycle.

A better metric is to track the total hours expended on the budget cycle. This requires salaried staff to track their hours, which they may not be accustomed to. The result shows the actual working hours spent on budgeting activities. A further refinement of this metric is to multiply the hours worked by the burdened pay level of the people charging hours to the budget. Since some of the participants are among the highest-paid members of management, this can provide an illuminating picture of the price of creating a budget, and particularly the price of creating extra iterations of the budget.

Another metric is the number of accounts listed in the budget. Since the budget is easier to construct when there are fewer accounts, this provides an indication of the simplicity of the budget model. The same approach can be applied to headcount—just measure the number of positions separately itemized in the budget.

These metrics provide a general view of the efficiency and duration of the budget cycle, which indicate management's success in budget compression.

SUMMARY

The budget process is unlike many of the other systems discussed in this book. It involves no transactions and does not require data entry or the transfer of assets. Because of these differences, the budget creation process has little need for controls. There are significant opportunities to streamline the process by avoiding an excessive number of iterations, shrinking the size of the model itself, and using a formal procedure to control the process pace. By implementing these changes, a quality budget can be produced more quickly than had previously been the case.

Chapter 9

Closing the Books[1]

In most companies, the aggregated financial results of the previous month should be available within the first few days after the end of the period. The resulting financial statements are being demanded by corporate management, outside investors, and the Securities and Exchange Commission (SEC) (for public companies) on the shortest possible timelines. However, the closing process has traditionally been a slow one—two weeks is a common closing interval, with even the best companies only reducing the interval to four business days. This chapter describes how to close the books and create financial statements much faster—even in a single day.

CURRENT SYSTEM

The closing process follows multiple parallel paths, one for each of the functional accounting areas: payroll, invoicing, payables, inventory, and cash. These are noted in Exhibit 9.1. These separate process flows also interact with each other; for example, nearly all accounts payable steps must be completed before one can roll new purchasing

---

[1] The information in this chapter is summarized with permission from *Fast Close*, Bragg, 2005.

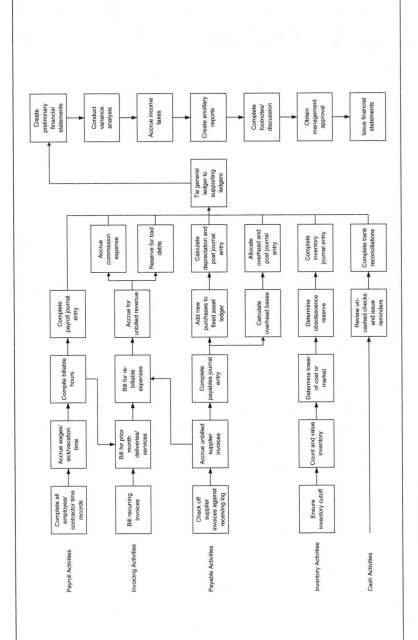

Exhibit 9.1 The Traditional Closing Process

information into the fixed asset ledger or the overhead cost pools for subsequent allocation. Because of these interdependencies, some processing steps can be significantly delayed. Also, controllers frequently wait while additional information arrives from outside sources before they are willing to complete a processing step. For example, they want to have a bank statement in hand before conducting a bank reconciliation; also, they may not close the accounts payable function until all significant supplier invoices have arrived.

The following bullets contain descriptions of the most common closing steps for each of the functional accounting areas:

- *Payroll activities.* If a company's payroll system does not produce paychecks on the last day of the month and include in those payments all time worked by employees through month-end, then some labor expenses are not being included in the financial statements. If so, the accounting staff typically spends a few days collecting time cards or similar documents from employees and calculating an accrued wage expense. They also must calculate the amount of accrued sick and vacation time and deduct from that figure any such time used by employees to see if accrual changes are required. Further, they must extract from the payroll records the amount of any billable hours and forward this information to the invoicing staff for billing to customers. Finally, they use the payroll register from any payrolls run during the month to create one or more payroll journal entries to transfer payroll information to the general ledger; this last step may be automatic if the company runs its payroll in-house, or is compiled from paper documents if the company uses an outside payroll processing provider. Of all these activities, controllers sometimes ignore the wage and sick/vacation expense accruals, which can render the financial statements significantly inaccurate if wages are a large proportion of total company expenses.

- *Invoicing activities.* If a company issues recurring invoices for such items as periodic maintenance contracts, it typically issues them at the beginning of the month. A services company will also use billable hours information from the payroll staff and re-billable expense information from the payables staff to create invoices for services rendered. A product company will use shipping information from the warehouse (usually transmitted on paper) to create invoices for recent shipments. If shipment information has a history of being inaccurate, an accounting or internal auditing person may also be sent to the shipping dock to manually review the shipping log to ensure that all shipped items have been billed. Also, if there are service hours earned but not billed, the accounting staff creates an accrued revenue journal entry. The controller also consults with the collections manager to update the reserve for bad debts, while accounting clerks create a spreadsheet listing commissions earned by each salesperson; the sales manager usually reviews this commission report and modifies it based on changes in commission splits, invoice allocations, and other criteria. The accounting staff modifies the commission report based on these changes and books a commission journal entry. Of all these activities, controllers tend to ignore regular updates to the reserve for bad debts, which places the company in the dangerous position of running out of reserves in a subsequent month and having to record an unusually large bad debt expense to rebuild its reserve.

- *Payable activities.* Though not a difficult area to complete, controllers have a tendency to delay the closing in order to receive every last supplier invoice related to the period being closed. The more careful ones do this by obtaining a copy of the receiving log from the warehouse and checking off all invoices against it, waiting until every received item has a matching invoice. Once completed and rolled into the general ledger, this information is used to populate the fixed asset ledger and calculate depreciation, which in turn is also posted to the general ledger. Also, the completed payables information is rolled into cost pools for allocation to inventory. Thus, the accounts payable area is a significant bottleneck for key downstream closing activities.

- *Inventory activities.* Inventory is the area that gives controllers the most heartburn, because there is a risk of considerable inaccuracy in the reported figures. To prevent this, they tend to take extra time to ensure that the month-end cutoff has been properly completed and the inventory physically counted and valued. These steps can take many days, and so (like accounts payable) can be a bottleneck in the closing process. Additional closing steps are writing off inventory based on the lower cost or market rule, as well as adjusting the reserve for obsolete inventory. These last two steps are ones frequently skipped by controllers, which is dangerous—recording large expenses for these items only at year-end can be quite a shock to senior management, and has led to the firing of more than a few controllers.

- *Cash activities.* Conducting a bank reconciliation seems like an easy closing activity, but tends to be considerably delayed because some controllers insist on waiting until a physical bank statement arrives in the mail. Others delay the reconciliation until some time after the close has been completed, which underreports the amount of bank service fees listed on the bank statement. An added step regularly followed by more conscientious controllers is to regularly review uncashed checks and contact the payees to remind them to submit the checks to a bank for payment.

- *Additional activities.* The primary closing activities were noted earlier in Exhibit 9.1. However, other steps are necessary that will be specific to the circumstances of individual companies. For example, if there is outstanding debt at month-end, someone must calculate the amount of *accrued interest expense*. Also, if a company has entered into a royalty arrangement in exchange for its use of product designs, patented processes, and so on, then the *royalty expense* must also be accrued. Further, there may be changes to the *equity accounts* caused by stock issuances, buybacks, or a company's recording of changes in its ownership interest in other entities, all of which require additional journal entries. Finally, a closing activity applicable to multiple functional areas is the detailed *account reconciliation*; this involves determining the exact contents of key accounts, such as accounts receivable and payable, prepaid expenses, and a variety of liability accounts. Listings of the contents of some accounts are commonly kept on electronic spreadsheets.

The corporate controller of a multi-division company will review the last section and then loudly opine that the real bottleneck operation is missing—the forwarding of completed financial information from each division, which can take many days. The problem is exacerbated if the division controllers have no sense of urgency regarding the close, and prefer to issue thoroughly reviewed and approved financial packages to the corporate accounting staff, sometimes many weeks after the reporting period. The problem is even worse if there are multiple levels of corporate reporting, so one company reports its results to its reporting parent, which reports to its parent, and so on. This can cause massive delays in the reporting process.

An additional issue for the multi-division company is mapping the charts of accounts (COA) of all the reporting divisions to the corporate COA. If done manually, there is a considerable risk of mapping inconsistency from month to month, which can play havoc with subsequent variance analysis tasks. This is a particular problem for companies owning disparate business entities whose operational requirements really do require variations from the corporate COA.

Another issue is the identification and elimination of intercompany transactions. This is a significant task for companies whose business operations are heavily integrated both upstream and downstream. For example, a tire manufacturer who owns the raw rubber source as well as retail tire outlets must eliminate from its reported financial results any transactions between the various operating divisions prior to the final sale to customers. This is a particular problem when the various divisions have free-standing accounting systems, since there is no automated approach for linking a sale on the books of one division to a receipt on the books of the division to whom the sale was made.

A final problem is analyzing and correcting the information supplied by the divisions. Since there are likely to be differences in the quality of information provided by each division, there is a reasonable chance that corporate-level variance analysis will uncover problems in the underlying transactions. Since the corporate accounting staff rarely has access to each division's individual transactions, it must forward investigation requests to the divisions and wait for them to conduct a review, which adds time to the closing process. Also, if a correcting entry is required, the corporate staff usually makes the change in the corporate books, which means that the corporate record of the division's results now vary from the division's in-house accounting records, requiring a periodic reconciliation to bring the two sets of books into alignment.

Thus, waiting for the financial results of subsidiaries, consolidating the information, eliminating inter-company transactions, and investigating and correcting problems in the consolidated results can add multiple weeks to the closing process—possibly several times more than is required for a single-location entity to report its financial results.

The publicly held company has considerable additional reporting tasks beyond those of a private firm. The SEC requires a very specific reporting format (as detailed in its Regulation S-X) for financial statements. In addition, the annual 10-K and quarterly 10-Q reports require considerable additional disclosures. These reports are subject to intense SEC and investor scrutiny, so companies tend to spend lengthy periods of time reviewing their contents prior to issuance. If the underlying financial statements have been completed relatively quickly, the importance of issuing accurate 10-K or 10-Q reports leads many companies to still issue these documents only at the last minute, thereby giving them more time for document review.

The earlier exhibit showed the basic closing tasks for a single-location company. Having just covered the additional reporting intricacies of multi-division and public companies, the additional closing steps required by these entities are noted in the flowchart in Exhibit 9.2.

Another way to look at the closing process is using a timeline in which tasks are divided into four-hour time blocks. An example is shown in Exhibit 9.3. The time periods shown here are only estimates, and will vary considerably by company. Also, it assumes a multi-location company that must spend extra time consolidating the results

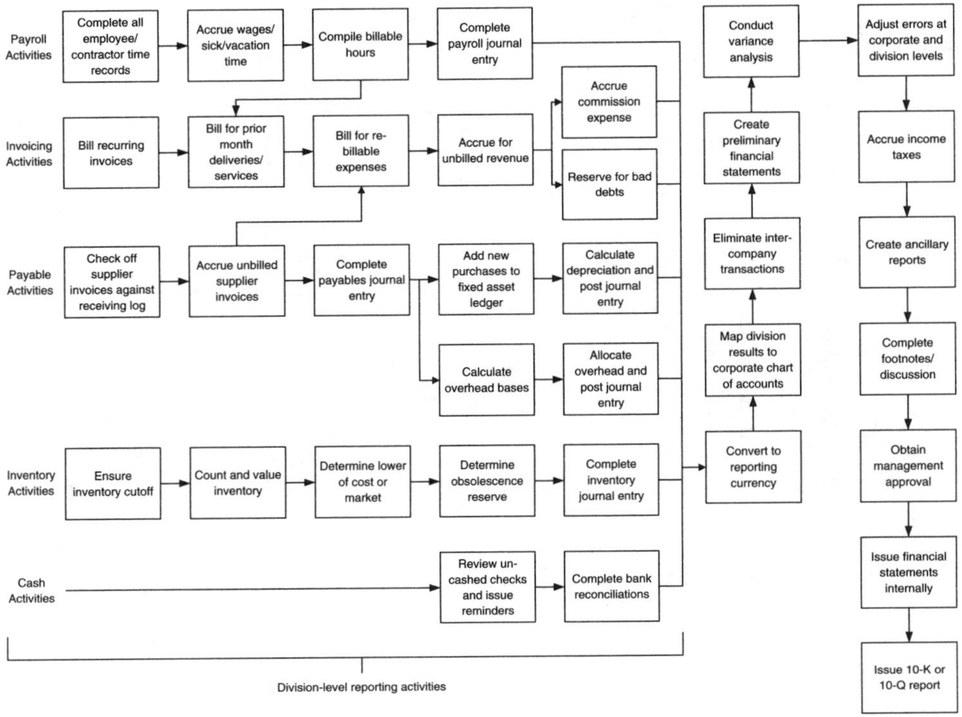

Exhibit 9.2 Closing Tasks for the Complex Enterprise

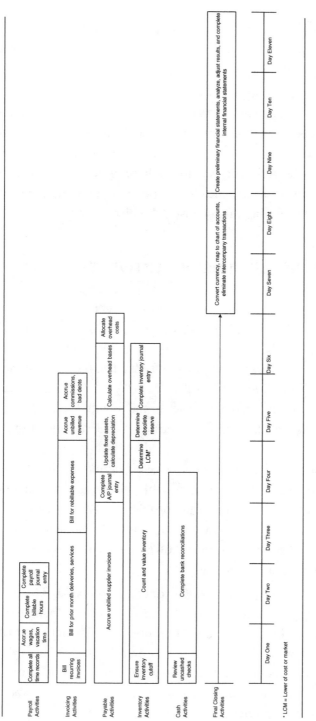

Exhibit 9.3 Combined Closing Timeline

279

of its subsidiaries. The exhibit reveals that a company can take 11 business days to complete its financial statements. We will use this format again later in the chapter to show how various improvements can shrink the closing interval.

The key point about the traditional closing process is that it initially appears to be a frighteningly complex and tangled mess through which the controller must painfully navigate—every month. However, as we will see in the following sections, there are a multitude of ways to more efficiently manage the closing process, resulting in a much faster close than the closing timeline would indicate is possible.

THE TIMING OF CLOSING ACTIVITIES

The closing process does not have to start on the first day of the closing month with a sudden rush of several dozen activities. Instead, many items can be partially or completely shifted out of this core closing period, either by moving them into the last few days of the preceding month or by recognizing that they are not central to the closing task, and deferring them until later in the month. Here are a number of activities that can be completed during less active parts of the month:

- *Review transactions.* The core closing period is a bad time to learn about transactional errors, since this requires a last-minute fire drill to determine the nature of the problem and derive a correction. Instead, review a preliminary set of financial statements a few days in advance of the close, which allows for the correction of these issues in a more leisurely manner.

- *Daily bank reconciliations.* Conducting a month-end bank reconciliation requires wading through a large number of bank transactions and painstakingly recording a number of adjusting entries. Instead, consider accessing the bank's online bank account information every day to conduct a smaller incremental reconciliation. By the end of the month, there should be only a minimal number of open items left to reconcile.

- *Update reserves.* Reserves are estimates, and so will not be any more accurate if they are calculated in the midst of the core closing period, or a few days sooner. In fact, by doing so sooner, there is more time available to fully analyze available information about reserves for such items as obsolete inventory and bad debts, so updating a reserve in advance may actually result in more justifiable reserve information than would otherwise be the case.

- *Bill recurring invoices.* The precise amount of a recurring invoice is usually known well in advance of the actual invoice printing date, so why not print it early? Just set the date of the accounting software forward to reflect the date on which they are supposed to be printed, which records the revenue in the proper month.

- *Review preliminary rebillable expenses.* Some expenses incurred by a company are rebillable to customers. If so, spend some time before the end of the month to review and organize this information, so that eventually billing it will be as trouble-free as possible. This review should include an examination of all expenses for receipts, and ensuring that expense totals add up correctly.

- *Review preliminary billable hours.* If a company issues a large proportion of its month-end billings on the basis of employee hours worked, the accounting department may find itself inundated during the core closing period with the detailed analysis of hours worked by a job prior to entering this information into invoices. A careful analysis of hours worked is usually necessary in order to avoid customer nonpayment of invoices for a variety of reasons: hours incorrectly coded to the wrong job, hours charged in excess of authorized funding limits, incorrect labor rates, and so on. To avoid these problems, review the information just before month-end.

- *Accrue interest expense.* If there is outstanding debt and interest payments not made each month, there should be a charge to interest expense that accurately reflects the amount outstanding throughout the month. Accountants typically wait until the month has closed before making this entry, in order to incorporate into the expense calculation any last-minute changes in the debt balance. However, if a change occurs within just a day or two of month-end, how much will this really change the total interest expense? Unless the amount of a last-minute change in debt is extremely large, completing the interest accrual a day or two before month-end will have little impact on the total interest expense charged.

- *Accrue unpaid wages.* If there are many employees who are paid on an hourly basis, the size of the month-end wage accrual can be substantial. If so, the accounting staff may spend a great deal of time during the core closing period, calculating every hour worked but not yet paid. However, it is possible to make a reasonable estimate of the accrual if the workforce is stable and consistently works the same number of hours each day. An alternative is to use a centralized timekeeping system to obtain the most current information about hours worked, and generate the accrual from that central database at the end of the day just prior to the core closing period. If not all of the timekeeping information is available at this point, the missing hours may constitute such a small proportion of the accrual that the accounting staff can reasonably estimate the size of the wages associated with the unrecorded hours.

- *Reconcile asset and liability accounts.* A prudent controller will insist on having a complete set of detailed account analyses as part of the financial statements. Doing so will ensure that no unusual assets or liabilities are parked on the balance sheet when they should have been flushed out through the income statement. However, it takes time to prepare this information, which can clog up the closing process. Instead, prepare the reconciliations on the last day of the preceding month. There is still a risk of having a few last-minute transactions dump items into these accounts, but those items can be added in the following month's reconciliation.

- *Calculate depreciation.* Depreciation is generally one of the last steps in the closing process, because it cannot be completed until the payables ledger is closed and used to update the fixed assets ledger. However, it is possible to develop a reasonably accurate depreciation expense a few days early. Doing so involves the risk that a fixed asset would be acquired or disposed of during the last day or two of the month. Even if some asset transactions are missed, is this of any real importance? Fixed assets are depreciated over many years, ranging from three years to more than a decade. Even assuming the shortest depreciation period of three years, the impact on depreciation of missing a month is $1/36^{th}$ of the total depreciable portion of the asset cost. Of course, the missing depreciation will still be recorded in the following month (thereby doubling the depreciation expense in a single month for that asset), so the total depreciation expense for the year will still be correct.

- *Compile preliminary commissions.* The commission expense is one of the most time-consuming and frankly irritating parts of the closing process, because it is subject to the whims of what may be a Byzantine commission plan, including all types of splits, allocations, bonus rates, and other adjustments. It also begins after all invoices are completed, so commission calculations are compressed into the end of the closing process, just when there is extreme pressure to complete the financial statements. Instead, consider beginning to close commission calculations a few days in advance. Presumably, some invoices were issued prior to month-end, so the commission on each one can easily be calculated in advance and verified with the sales manager. By doing so, only invoices issued at the last minute still require a commission calculation, thereby greatly reducing the volume of work to be completed on closing day.

Many closing tasks can be completely or partially shifted out of the core closing period. This is a simple yet effective way to create rapid improvements in the duration of the closing period. In addition, by completing some tasks during the less-frenzied period a few days prior to the close, the accounting staff may find that it has the leisure to do a better job of transactional analysis, and so creates fewer errors that would otherwise find their way into the financial statements. As shown in Exhibit 9.4, the impact of these changes on the total duration of the close can be profound, with three days of effort disappearing from the

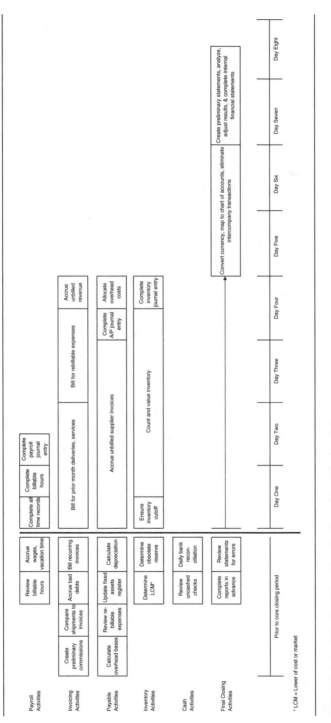

Exhibit 9.4 Closing Timeline Reduced by Timing Changes

* LCM = Lower of cost or market

core close, and shifting forward into the days preceding the end of the month.

REDUCE THE CONTENTS OF THE FINANCIAL STATEMENTS

The typical financial statement package is too large. It includes lots of operational and financial information beyond the customary set of financial statements. While some of this information, such as subsidiary and departmental financials, can be readily printed from the accounting system with minimal delay, other reports require a great deal of extra preparatory work, and so lengthen the close.

It is useful to see which pages of the financial statements are being read; if they are of minimal use, then strip them out. This concept can extend to individual line items, such as operational measurements, that require a great deal of time to prepare. To determine the efficacy of this information, conduct a periodic survey of managers, asking them what information they use. Always conduct this survey in person, in order to extract the highest-quality responses.

Some departments or divisions like to see their financial statements in a particular layout and may have prevailed upon the controller to provide them with this format. The result is continual restatement of the financials into several different reporting packages, which require extra time to prepare. Instead, work toward the use of just a single reporting package for everyone, or at least the smallest possible number of variations from the basic package. If this is not possible, then a more complicated approach is to create an intranet portal containing a variety of reporting templates that are linked to a data warehouse in which the financial statements are stored. Authorized users can then use any report template they want in order to download reports in the precise configuration they like to see.

The most time-consuming and error-prone part of the reporting package is costing reports. It frequently requires detailed financial analysis in the midst of the core reporting period, and may require reconciliation to the financial statements, or be disputed by managers. To avoid these problems, recast the cost reports into a separate reporting package that is issued well away from the financial reporting package. Managers will be most amenable to this approach if the cost reports are issued more frequently, perhaps weekly. By doing so, the events causing variances will have only just occurred, making it easier to fix their causes on an ongoing basis.

Creating metrics for the financial reporting package can add hours of work to the closing process, because metrics usually require some manual data gathering or calculations. This can be avoided by separating the metrics from the standard financial statements, perhaps issuing them a day or so later. An alternative is to still issue metrics within the standard reporting package, but only those that can be automatically generated.

In short, paring down the contents of the financial statements package will reduce the effort required to issue the financial statements. This is an activity that should be repeated at least once a year, since additional reports and line items have a way of creeping back into the financials package on an ongoing basis.

CLEAN UP AND STANDARDIZE JOURNAL ENTRIES

A picky controller who wants perfect financial statements has a tendency to create far too many journal entries to polish up every account, sometimes for extremely small amounts that most people would consider well below the lower limits of materiality. Small journal entries also arise when they were originally created to handle much larger dollar amounts, and were never eliminated when the reasons for the entries shrank in size. In either case, immaterial journal entries waste time during the closing process and should be eliminated.

The decision to use a journal entry is quite simple: when in doubt, leave it out. In other words, there must be a clear and discernible improvement in the financial statements as the result of a journal entry.

This approach also applies to the use of accruals to park smaller unrecognized expenses in prepaid asset accounts. It takes time to create the journal entry, prepare a detailed analysis of the asset account, and reverse the entry in a later period when the expense needs to be recognized. In exchange for the immediate recognition of the expense, avoid all these steps by not making the journal entry.

There can easily be dozens or even hundreds of journal entries created during a single reporting period. With so many being created, it is difficult to wade through them all to see if the key entries are being made every month and if they are using the same accounts from period to period. If not, the reported financial results will be extremely inconsistent. For example, if a wage accrual entry is only made sporadically, the reported level of wage expense will increase when the entry is used and decline when it is not.

To fix this problem, begin with a standard checklist of journal entries to be completed as part of the closing. The checklist should include a check-off box or space for initials to indicate completion, the approximate timing of the entries, and a reference number to show the name of the journal entry template as it is stored in the accounting system. An example of such a checklist is shown in Exhibit 9.5.

Exhibit 9.5   Journal Entry Checklist

Initials	Timing	Journal Entry	Reference Number
SB	3 days prior	Inventory obsolescence reserve	IN01
SB	3 days prior	Lower of cost or market	IN02
SB	2 days prior	Bad debt reserve	AR01
SB	2 days prior	Interest expense	GL01
SB	2 days prior	Unpaid wages	PR01
SB	2 days prior	Unused vacation time	PR02
SB	1 day prior	Bank reconciliation	CA01
SB	1 day prior	Unbilled supplier invoices	AP01
SB	1 day prior	Overhead allocation	IN03
SB	1 day prior	Depreciation expense	AP02
SB	Closing day	Unbilled revenue	AR02
SB	Closing day	Commission expense	PR03
SB	Closing day	Royalty expense	AP03
SB	Closing day	Income tax liability	GL04

The reference number shown in Exhibit 9.5 uses a two-letter designation for the functional area most closely related to the journal entry, plus a number to indicate the volume of templates within that functional area. For example, the first reference number in the table is IN01, which indicates the first inventory journal entry. Similarly, the last entry in the table is designated GL04, which is the fourth general ledger entry. Some computer systems allow for very large file names, so less cryptic designations can be used.

As just noted, a key part of journal entry standardization is the use of a template. This is a blank journal entry kept in the accounting software, which may be called a template, format, or memorized transaction, depending on the software being used. Users simply go to the journal entry checklist to find the stored name of each template, enter numbers in the template, and save it in the correct accounting period. Templates are intended for any journal entries used on a repetitive basis for which the account numbers stay the same but the dollar amounts of the entries vary. One should occasionally review the templates to verify that the account numbers listed within them are still correct—a quarterly review is sufficient.

Most closing processes require the use of some journal entries, and frequently a great many of them. Because some of these entries can be complex, there is a high likelihood that some account numbers or dollar figures will be entered incorrectly. Another very common problem is transposing entries, so debits are recorded as credits, and vice versa. These errors are difficult to spot, and always time-consuming to correct. If not found, they can have a significant impact on reported financial results.

One solution is to use the recurring journal entry feature in the accounting software. This allows users to record a journal entry, with all account numbers and dollar figures, and then state the number of accounting periods over which it shall be in effect. The system will take matters from there, automatically recording journal entries in succeeding periods until entries have been recorded for the full range of designated months. In the interim, the accounting staff has no work to do, which can make a small dent in the length of the closing process.

The recurring entry is best used for transactions that have no chance of requiring adjustments over the period when they are pre-loaded to run, such as the amortization of a specific value for a designated period. It can also be used for large journal entries where some changes are likely each month, but revising the recurring entry is still less time-consuming than using a template. An example is a depreciation expense entry where there are continuing updates to dollar values in the entry based on asset additions and dispositions.

When using recurring entries, it is useful to create a printed list of all such entries and compare it to the entries appearing in the general ledger each month. This step ensures that all recurring entries are running that are supposed to run. Also, consider inspecting in detail any recurring entry that is in its final month of activation. These entries may require slight adjustments so that the total value of a series of recurring entries matches the goal amount. For example, if the objective of a series of recurring entries was to amortize an initial value of $12,003 over 12 months, the monthly entry may have been $1,000, which leaves an extra $3 to adjust in the final recurring entry in order to completely eliminate the initial value.

It is not especially difficult to learn how to create a journal entry in a company's accounting software. This can have the same impact as giving the family car to a teenager with a learner's permit—havoc can ensue. The problem is that more than one person may create a journal entry for the same transaction, resulting in duplicate entries or (if there is no adequate standard procedure in place for a closing activity) a cluster of slightly different entries. When this happens, someone must spend time reviewing the journal entry list for duplications, determine which one is correct, and delete the others. Also, if a person making a journal entry does not have adequate accounting training, the entry may be to the wrong accounts or have flipped the debits and credits.

To keep these problems from occurring, one well-trained person should be designated the general ledger accountant, and be solely responsible for all journal entries. If the computer system allows it, consider using passwords to lock out all other users from journal entry activity.

The net effect of cleaning up and standardizing journal entries is a small improvement in the speed of closing, but can be substantially greater if the original closing process included a large quantity of unregulated journal entries.

STANDARDIZE THE CHART OF ACCOUNTS

A major cause of delays and data entry errors towards the end of the closing process is the mapping of information from each subsidiary's chart of accounts (COA) into the master COA used by corporate headquarters. The differences between the various COAs can be substantial, making the data transference chore a major one. Some companies accept the differences between COAs and create complex automated consolidation mapping systems, but there is still a risk that a mapping from one subsidiary account to a corporate account will be set up incorrectly, creating incorrect financial results. Also, if subsidiaries are allowed to change their COAs at any time, a considerable effort will be needed to continually ensure that the automated consolidation mapping system is keeping up with these changes.

The solution is to adopt a standardized COA for the entire company. This means visiting all the subsidiaries to determine the need for any variations from a master COA, incorporate these changes if needed, and then creating a company-wide policy to lock down all COAs, with subsequent changes allowed only with permission from the corporate staff. By doing so, the account mapping task is completely eliminated from the list of closing activities, as are any attendant mapping errors. *This is a major improvement for the closing process.*

Standardizing the COA can be particularly difficult to implement if the subsidiary accounting divisions have considerable autonomy, or if the baseline COA is extremely large (requiring great effort to pare down). In either case, a good long-term solution is to implement a single enterprise resource planning (ERP) system that is used throughout the company, and where access to the COA is locked down and only accessible by the corporate accounting staff.

CENTRALIZE ACCOUNTING FUNCTIONS

Controllers who have the worst trouble shrinking the length of the closing period are probably those whose companies have many people involved in the process, using different procedures in different locations. These two issues are an insidious and significant reason why some companies never achieve a fast close, because the inherent level of procedural complexity is too high. In this section, we will review the need for standardization and centralization within the closing function, and how the implementation of these two concepts will contribute to a faster closing process.

One of the most prevalent reasons for a slow close is that the corporate accounting staff must investigate variances and outright errors in the summarized information they receive from divisional accounting staff, and spend days correcting them. Frequently, a problem is caused by a multitude of different accounting procedures throughout the company, resulting in inconsistent reporting results that are hard to reconcile. Also, since different divisions use different accounting systems, it is impossible to spread accounting best practices throughout the company—they may work well in one location, but not in another.

The solution is to implement a common approach to all accounting transactions throughout the company, not just to those transactions directly related to the closing process. This approach improves the level of corporate control over transactions. Further, if a recurring transaction error is discovered that presents the risk of significant and ongoing reporting errors, the corporate controller can easily mandate a change in the standard accounting procedures, thereby rolling out critical system changes in a short time period. This approach improves the level of information accuracy at the source, resulting in far less time being required by the corporate accounting staff to track down and fix errors at the end of the closing process.

By standardizing accounting systems throughout the company, it also becomes much easier to mandate the use of closing schedules and force the accounting departments of subsidiaries to follow them. For example, if a division controller claims that he cannot forward closing information by a scheduled date, she can no longer use as an excuse the presence of a unique step in her divisional closing process—such steps are not allowed. Also, the corporate controller can more easily judge the management talent of the divisional accounting staffs based on their ability to produce financial results in accordance with the closing schedule, since everyone operates under the same guidelines.

Further, the use of standardization eliminates the need for information being rolled in from unrelated sub-systems or electronic spreadsheets, which can be time-consuming and is more likely to include errors. Instead, the common closing procedure specifies exactly where all data comes from, how it is processed, and where it is posted as part of the closing process. There is no longer a place for unusual sources of information. However, since divisional controllers may adopt their own systems to process data, it is important to dispatch internal auditors to the divisional accounting departments from time to time in order to search for and report on the presence of such systems.

No matter how much standardization is incorporated into the closing process, there will still be labor inefficiencies as well as an increased level of errors if accounting transactions are processed in a large number of company locations. This is due to the inherent difficulty of running operations in multiple locations, the increased number of work queues caused by the involvement of more people, and variations in accounting procedures amongst locations.

One of the best ways to resolve the closing problems caused by accounting decentralization is to require all accounting transactions to be processed from a central location. By doing so, specific types of transactions (e.g., accounts payable, billings, general ledger) can be concentrated along functional lines with a smaller number of more highly trained people, resulting in fewer errors. Also, by regrouping responsibility for accounting tasks along functional lines rather than geographical ones, it is easier to assign responsibility for various closing tasks to a smaller number of managers, making the closing process easier to monitor.

An added advantage of a centralized accounting system is the ease with which accounting errors can be researched, which reduces the time required to resolve variances in the financial statements. The improvement results from having a single central database of accounting information, allowing financial analysts to more easily drill down through layers of data to locate problem areas. This is a great improvement over a dispersed accounting operation, where the analysts are forced to ask outlying accounting departments to research issues for them, for which they may wait days to receive an answer. Thus, access to centralized accounting data eliminates the wait time associated with putting variance analysis in the work queues of divisional accounting departments.

The advantages of centralization that have just been enumerated will not be fully realized if the corporate accounting department does not use a single consolidated accounting system. If there are multiple systems at corporate headquarters requiring manual interfaces, then the drill down capability will be severely restricted. Also, different systems have a varying "look and feel," making it more difficult to train new employees in their use. Further, software updates to one accounting system may crash any customized interfaces constructed to more easily swap data between systems. For all these reasons, a single consolidated accounting system is the best foundation for a centralized accounting system.

Though the improvement caused by centralization will certainly be noticeable in a multi-location company, it also has an impact within a single-location organization.

In a normal accounting department, closing tasks are handed off from one person to another, and yet another. At each handoff, the closing process is lengthened by the wait queue of the person receiving the work. Thus, if tasks must pass through many people, the closing process will be inordinately long. The problem with wait queues can be reduced through a concept called process centering. Under this approach, the controller should reduce the number of people involved in the closing, thereby eliminating time wasted while information queues up between people, as well as the risk of work not being done if someone is absent. The decision to centralize closing tasks with fewer people is based on an ongoing analysis of wait times within the closing process; the controller focuses on eliminating from closing tasks any positions creating bottlenecks in the process.

The standardization and centralization of accounting activities has a major positive impact on multi-division companies having local accounting departments. As a result of the standardization and centralization actions noted here, the time required to complete final closing activities can drop from multiple days to just a single day.

INVENTORY TRACKING

A major bottleneck in the closing process is counting inventory at the end of each month. A physical inventory count can be eliminated if accurate perpetual inventory records are available. The complete set of steps needed to create a perpetual inventory tracking system was already described in Chapter 4, Inventory. If completed and properly maintained, the accounting staff can rely upon the computerized database of inventory records to automatically value the inventory. The net result of this change is the elimination of inventory as a bottleneck from the closing process.

PAYABLES PROCESSING TIME COMPRESSION

The one remaining functional area that can be difficult to close quickly is accounts payable. Some controllers fear that a late-arriving supplier invoice will drastically alter the results of a month, and so they wait days for all invoices to arrive before finally closing the month.

There are several ways to address this problem. One is the mandatory use of purchase orders for all larger purchases. By doing so, the company has a record in its purchasing database of all outstanding orders of any size. The accounting staff can access this information to accrue for any missing invoices, thereby eliminating all uncertainty from the process.

If there is no purchase order system, then the accounting staff can instead compare the receiving department's receiving log against its supplier invoice records. If there is a receipt with no corresponding invoice, then an expense accrual is in order.

Yet another approach is anticipation of expenses based on historical information. If an invoice from a specific supplier is always late and is reasonably predictable in size, then accrue the historical average amount for that invoice. This approach is the weakest of the three alternatives, since it provides no safety net for large, unexpected supplier invoices.

Any one or combination of these three methods should allow the accounting staff to close the payables function very quickly, eliminating it as a bottleneck in the closing process.

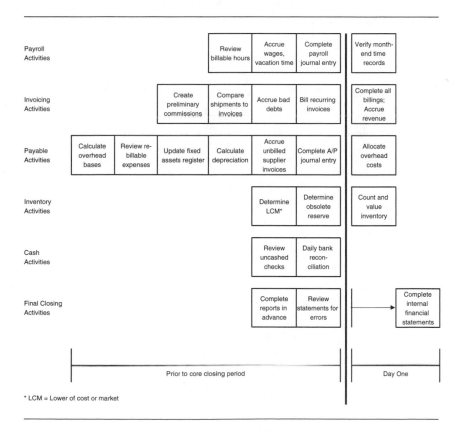

Exhibit 9.6   Closing Timeline Following All Improvements

SUMMARY

In summary, a variety of techniques are available that allow a company to close an accounting period much more quickly than before. No single change described here will trigger a massive reduction in the closing interval. Only by gradually working through the majority of these improvements will a company eliminate enough bottlenecks to arrive at a fast close. The result is shown in Exhibit 9.6, where nearly all of the closing activities are now completed prior to the core closing period, with only a few residual events occurring after the end of the month. Generally speaking, the main remaining transactional event after month-end is customer billings.

For a more comprehensive treatment of this subject, see the author's *Fast Close* book.

Chapter 10

Data Collection and Storage Systems

One of the most time-consuming aspects of the accounting function is the collection and storage of data. In this chapter, we will review some examples of traditional systems as background for the discussion of more advanced systems that can dramatically cut the time required for these functions.

TRADITIONAL DATA COLLECTION AND STORAGE METHODS

In most accounting environments, the number one data collection area is payroll. One of the more tedious tasks for the payroll clerk is the accumulation of pay data from each person who is paid on an hourly basis. This information must be checked for accuracy, approved by a supervisor, summarized, and entered into the payroll calculation software so that it can be converted into paychecks. These data collection tasks occupy the bulk of the payroll clerk's working time—the remaining tasks are minor in comparison. This is a particularly laborious and expensive task if the hourly work force is large. There is an enormous time crunch at the end of each pay period, when all of the time records must be examined within a very brief time frame. This problem is further exacerbated if each employee charges time to specific jobs, because this additional information must be verified against a list of open jobs, approved by a supervisor, and keypunched into a job costing database.

Another area in which data collection represents a significant proportion of the total time spent is cost accounting. The cost accountant is constantly sorting through records to determine the cost of a job. This can include labor costs, as just noted, plus the unit volumes of supplies and materials charged to a job, as well as the time that each item spent on a particular machine. This can be an exceptionally complicated cluster of records to assemble, and usually requires a number of cross-checks and iterations to ensure that the data is both accurate and completely represented.

A continuing data collection problem area is in the three-way matching of supplier invoices, receiving documentation, and purchase orders. This requires the collection of data from three different parts of the company (e.g., the mail room for supplier invoices, the receiving dock for receiving documentation, and the purchasing department for purchase orders). To have all three types of data come together without any extra investigative work is the exception rather than the rule.

Another area of data collection is the billing function. The shipping department must create a bill of lading and forward it to the accounting department, which matches it with a customer order (which either comes from the order entry department or is attached to the bill of lading), compares the information to a standard price list, and creates an invoice. If the price on the order does not match the standard price, then the accounting staff must track down the salesperson who took the order to ascertain the correct price. Also, the list of outstanding customer orders must be compared to the orders invoiced, to see if anything has been shipped that has not been billed. If there are exceptions, then this issue also requires investigation.

Thus, data collection is a pervasive part of the accounting process, ranging from the direct collection of data in the payroll and cost accounting functions to the verification of data that is forwarded from other departments, as is the case for the accounts payable and billing functions.

Similar issues arise at the other end of the accounting process, where data must be properly indexed and stored. Payroll records are a particularly thorny problem because government regulations require that they be kept available for a number of years. Time cards are usually batched by week and kept in storage boxes. Though this may seem like a reasonable approach, even a single search for records will require pulling boxes from storage, rooting through them to find the correct week, and then finding the correct time card. Putting all of this back together is another task that requires time and care.

The data storage problem is less of an issue for accounts payable and receivables, since these are usually filed in separate folders for each supplier and customer. Nonetheless, repeated refiling of documents within each folder will also mix the documents by date, rendering subsequent document file searches more difficult.

Perhaps the worst data storage problem is presented by the cost accounting function. Most companies do a notoriously poor job of storing this information, for they see no reason to keep it for more than a few weeks after the completion of manufacturing. However, they are incorrect in this assumption because cost records are an important part of both the year-end audit and of any government contracting work. An external or government auditor has a keen interest in these records, since they can be used to verify the accuracy of year-end inventory records, as well as of government billings. It can be quite a chore for the accounting staff to find the correct records when they have been unceremoniously dumped into a storage box with no further identification on the box than the year to which the records relate.

The data collection and storage problems described here are not those of the rare slipshod company—these are typical systems for most organizations. The systems described in the next section reveal the broad array of improvements that technology can bring to the data collection and storage methods of any company.

ADVANCED DATA COLLECTION AND STORAGE METHODS

A variety of technologies now make it possible to eliminate a large proportion of a company's data collection tasks. These technologies include the use of computerized time clocks, interactive voice response systems, bar coding, digital pens, and more. These alternatives are described below.

COMPUTERIZED TIME CLOCKS

The previously described problems with payroll data collection can be almost completely eliminated through the use of a new breed of electronic time clock. These can be found in three forms—those that record employee time based on bar-coded employee cards, magnetic-stripe cards, and biometric measurements of an employee's hand.

The first of these clocks requires a person to slide an employee badge through a slot on its side that reads a bar code that is imprinted along the edge of the badge. This bar code contains the identifying number of each employee. The clock then stores the date and time when the card swipe took place. When the employee is ready to leave work at the end of the day, he or she scans the badge through the time clock again, which stores this information yet again. The data stored in the time clock is usually downloaded on a daily basis to a central payroll computer. The application software in this computer then compares the two scans to automatically determine how much time the employee worked, which eliminates the chore of manually calculating hours worked.

The second of these clocks operates in exactly the same manner as the first, except that it reads an employee badge that contains a magnetic stripe, rather than a bar code. This type of card is usually purchased from the time clock supplier, as opposed to the bar-coded variety, which can be readily assembled on-site. Many companies choose to use the magnetic stripe version, on the grounds that bar codes are easily duplicated on any photocopier, whereas duplication is much more difficult with a magnetic stripe. This can be an important issue if an employee chooses to create multiple copies of an employee badge and give it to co-workers, so that they can scan that person into the timekeeping system, even if that person is not on the premises.

The third type of clock uses biometric analysis to virtually eliminate the risk of having employees falsely clock into the payroll system. This clock records the precise outline of a person's hand and stores this information, so that it can readily identify employees at any time they place a hand over the clock's scanning pad. Though a more expensive option than the previous two, this system eliminates the cost that would otherwise arise from paying employees for work not performed.

There are a number of important additional features that all three clocks contain. One is a standard report that is issued from the central payroll computer that compares the record of all employees who have made an entry into the system to a list of those people who are supposed to be working—the difference is the name of every person who did not show up for work. This report goes to the production supervisor, who can use it to contact missing employees. Another key feature is that it can lock out employees to prevent their making an entry into the system either too early or too late; by doing so, it is not possible for employees to be paid for hours worked before their scheduled start times. Anyone arriving late must find a supervisor with an override code who can enter them into the system. Yet another feature is a report that itemizes all missing scans, so that the production supervisor will be aware of anyone who clocked in, but not out, or vice versa. It can also automatically log employees in and out of the payroll system when they are scheduled for meal breaks. More advanced versions of these clocks that contain keypads also allow employees to punch in the identification numbers of jobs on which they are working, which is useful if a detailed job costing system is in place.

Despite the numerous advantages of these devices, they are expensive—up to $2,000 each. Furthermore outlying clocks may require a phone line or Internet access in order to download their data to a central computer, which calls for an additional investment in cabling.

Offsetting these costs are the near-total elimination of time card calculations by a payroll clerk (though missing scans must still be investigated), which can be a substantial cost if there are many employees who are paid on an hourly basis.

There are two common pitfalls associated with computerized time clocks. First, management installs too few of them, on the grounds that they are so expensive. This is a mistake, since employees must waste time queuing up in front of the few operating units in order to clock in or out. This is also a problem if employees must travel to the far side of a production facility to reach a clock—it is better to have them conveniently placed near all major points of entry and exit. The other problem is that employees don't see a physical record of their time worked. The new system merely stores it in the computer database, where employees cannot see it. To allay employee nervousness over this issue, have formal training sessions with employees to show them how the system works, how the data is stored, and what safeguards are in place to ensure that their payroll information cannot be lost.

TIMEKEEPING BY PHONE

An alternative to the time clock for payroll data collection is the telephone. Under this approach, the company buys a rack-mounted server that contains an interactive voice response (IVR) system, and links it to their phone system. Employees then call into the IVR system to enter their time in response to a series of prompts. The capacity of the system ranges from one employee to over 100,000.

A demonstration IVR system has been set up by Telliris, Inc. To use it, call 203-924-7000, extension 5000 and enter partner code 0000. Then use employee number 00001 to enter a variety of transactions, such as clocking in and out, reporting sick time, vacation time, bereavement, jury duty, and family illness.

A timekeeping IVR system requires a reduced investment, since it takes advantage of existing phones. Also, the system is so intuitive that employee training is minimal. Furthermore, the system has built-in validation, to avoid initial data entry errors by employees. It is even possible to limit phone calls to specific telephone numbers (presumably originating from fixed phone locations), so that employees can only call in from where they are supposed to be. This is a very good solution for mobile employees, such as those involved with equipment servicing, facilities maintenance, and health care. It is also useful for temporary employees, since the company would otherwise have to issue them an employee badge in order to use any in-house timekeeping systems.

In addition to Telliris, timekeeping IVR systems are also offered by TimeLink and TALX.

BAR CODING

One of the data-gathering options noted above for timekeeping was a time clock that accepted bar-coded employee badges. The same concept can be used to gather other forms of data, except that the scanner used does not have to be in a fixed location, as was the case with a time clock.

A typical application for a bar code scanner is in the tracking of fixed assets. Whenever an asset is purchased, a bar code is affixed to it. Then, when the accounting staff conducts a periodic audit of fixed assets, an employee will be given a portable bar code scanner with internal storage; this person walks through the facility, scanning each bar code encountered. The scanner is then taken to a networked or central computer, where its contents are downloaded and compared to a master list of all recorded fixed assets, to see which ones are still not accounted for. A more advanced version of this scanner, the radio frequency scanner, will transmit scanned data back to a central computing facility in real time, so that assets can be compared to a master list at once. The central computer can then transmit back a list of missing assets, so that the person doing the scanning can conduct a search on the spot. This concept can be taken a step further, so that the person conducting the audit can enter into the scanner any revised asset locations, which will update existing asset locations in the central database.

The most common application for bar codes is in the area of inventory management. Under this approach, a bar code that contains the identifying part number for each inventory item is affixed to it at the receiving dock and immediately scanned, which records the receipt in the inventory database. The bar code is then scanned as it progresses through each stage of the production process, so there is an ongoing record in the production database that precisely identifies the location of each item. This is of great use to the cost accountant, who has a much easier time determining the cost of each item in the production process, as well as the overall cost of the inventory. This is also useful for the materials management staff, which can use this information to locate materials within the facility. An alternative use for bar codes in this application is to attach bar codes only to products at the *end* of the production process, so they can be scanned when a product is completed; this can trigger a backflushing transaction that will remove all materials from the inventory database that were used to create the finished product. Either of these alternatives nearly eliminates the need for the manual entry of transactions throughout the production process.

Before installing a bar coding system in the production area, there are several issues to be aware of. First, the resulting data will not be accurate if the part or product codes encoded into the bar codes are the wrong ones. This problem can occur quite easily, and will result in continually incorrect scans from that point forward, since it is highly unlikely that the incorrect bar code will be spotted, unless an alphanumeric code is also printed beneath the bar code and is regularly compared to the item being scanned. Second, bar codes are not always scanned as they progress through the production facility. For example, a materials handler may forget to scan raw materials when moving them from the warehouse to the production area, which will result in an incorrectly high balance in the raw materials account. Though this issue can be resolved through the use of automated scanners that will scan everything passing through a fixed point, these installations can be quite expensive. Yet another problem is dealing with exceptions. For example, if some parts within a bar-coded pallet are discovered to be faulty and are removed from the pallet, the bar coded quantity on the pallet must also be changed, while another bar code must be created that will assist in tracking the faulty products as they move through separate repair or return processes. This can result in a complex web of interdependent data-tracking processes—so complex that some errors will be bound to creep into the process, and which will require the ongoing services of cycle counters and internal auditors, who constantly review the process to check for holes in the system. It may also become an impossible system to maintain if the production staff that is responsible for all of these data collection activities is so poorly trained, or has such a high turnover rate, that they are incapable of consistently collecting accurate data.

The cost of scanning equipment must also be considered. For example, a single slot scanner or light pen that is positioned next to a computer keyboard can be acquired for as little as $100, but is also of limited use because it cannot be moved, and also provides only a single scan of a bar code, which frequently results in a missed scan. A better alternative is a moving-beam scanner, which will scan each bar code at a rate of about 30 times per second, and which will nearly always result in a correct scan; this type of scanner will cost anywhere from $300 to $600. If the intention is to make the scanner portable, then it becomes a self-contained and independently powered computer, which raises its cost to $1,000 to $2,000 for each unit. If a company also wants to give a portable unit the ability to communicate in real time via radio frequency, then the cost per unit can reach (and exceed) $4,000. If scanners are to be installed in a fixed location, with an expected scanning success rate of near 100%, then the cost can be much higher depending on the configuration of the installation. In addition, specialized bar code printers must be positioned wherever labels are needed, with each one costing at least $1,000, and much more if the required labels are exceptionally large or require special paper. Thus, a bar code scanning installation of even modest dimensions requires a significant capital budget.

Given the difficulties associated with installing a bar coded inventory tracking system, it is best to start with a small pilot project to ensure that the concept will work within your facility, and then expand the concept into those parts of the company where there is a minimum risk of incorrect data. This may result in a combination of automated and manual data collection systems, which is acceptable as long as it results in the optimal combination of accurate data and minimal data collection costs.

RADIO FREQUENCY IDENTIFICATION

Radio Frequency Identification (RFID) requires the attachment of a tiny transceiver to an object, allowing it to be tracked by a network of receivers. The most obvious use for RFID is inventory tracking at the pallet or case level within a business location, since this allows for a tightly-confined tracking area, moderate RFID cost, and tracking of what is frequently a company's largest-dollar asset.

The primary difficulties now involve cost (about $0.20 for a single RFID tag) and poor scan rates by receivers (currently about 70%). Though the technology clearly requires improvement, we should consider other applications once the problems are eliminated.

It is possible to use RFID in the accounting department. One possibility is the tracking of documents. Though many accounting departments already have document scanning systems in place that would appear to render RFID unnecessary, consider again—what about truly sensitive documents, such as signed legal documents, internal audit work papers, insurance folders, archived documents, and the like? In many cases, employees remove these documents from their customary locations and leave them all over the accounting department, making it more difficult to locate when they are needed on short notice. Perhaps adding RFID tags to these types of documents and adding receivers in the accounting area would be a reasonable way to ensure that they can always be found.

ELECTRONIC DATA INTERCHANGE

The accounts payable staff must re-key into its computers much of the information on every supplier invoice received. At a minimum, this will include the supplier's identification number, the invoice date, and the total amount due. A more comprehensive re-keying effort will also include the authorizing purchase order number for each line item on the invoice (or at least in total), as well as the cost and identifying number of each item purchased, plus the sales tax and freight costs to be paid. All of these entries require a great deal of time and will involve some degree of data entry error. To avoid these problems, have suppliers send their invoices to the company by electronic data interchange (EDI).

To use EDI, the sender must convert a transaction into a highly standardized electronic format (of which there are well over a hundred), and then send it as an electronic message to the recipient. This transmission can go straight to the receiving party, or (more commonly) be routed through a third party, where the message resides in an electronic mailbox. The recipient periodically polls the mailbox to extract any messages.

The key to the EDI system is to ensure that incoming messages are connected to an automatic interface to the accounting system, so there is no need to manually re-key the data upon receipt. By using this interface, the entire data entry task is eliminated. This system can be used for a variety of transactions, such as purchase orders that are sent to suppliers and invoices that come back from suppliers.

Though EDI can eliminate a great deal of data entry work, it is a difficult system to create, for it requires the cooperation of a large proportion of a company's business partners, each of whom must learn how to send and receive EDI transactions. They may be willing to do this if the company's business is significant, but less so if not. In the latter case, an alternative is to receive all paper-based invoices from suppliers and re-mail them to a data entry service that re-keys them into EDI format and then transmits the invoices by EDI to the company. The process can be made slightly more efficient by having suppliers mail their invoices to a lockbox that is directly accessed by the data entry service, so there is no need to re-mail invoices to the supplier. Though this may seem like an additional cost to the company, it actually removes from the accounts payable department *any* need for a secondary system that requires data entry—this may allow for the elimination of clerical positions, which can pay for the added cost of the data entry service.

DIRECT DATA ENTRY THROUGH A WEB SITE

A company may be experiencing some difficulty in persuading its suppliers to switch over to the transmission of invoices by EDI, which would allow it to automatically process all incoming invoices without any data re-keying. A typical complaint when this request is made is that special EDI software must be purchased and stored in a separate computer, while someone must be trained, not only in how to use the software, but also in how to re-format the invoicing data into the format used by the EDI transaction.

This problem can be partially avoided by having suppliers access a Web site where they can conduct the data entry. This allows suppliers to avoid the need for any special software that would otherwise be needed to generate an EDI transaction. A Web site merely requires Internet access. Once the data has been entered at the Web site, a company can port it straight into the accounting system.

There are some costs associated with this alternative. For example, a company may have to use special discounts or early payment enticements to convince suppliers to use the Web site, rather than simply mailing in their invoices. Also, the Web site must be constructed and maintained, while other software must be created that ports the incoming transactions into the accounting system for further processing.

DIGITAL PENS

A key goal of the accounting department is to convert all information into a digital format, so that it can be accessed through a computer, rather than a filing cabinet. Some types of data collection have proven to be quite resistant to change, especially where it is difficult to introduce a computer at the point of data collection. This is a significant issue when the sales staff is writing down sales orders in the field, new employees are filling out W-4 forms, internal auditors are working through control checklists, and so on.

A solution to these data entry problems is the digital pen. Made by Logitech, the io2 Digital Pen functions like a normal pen, but also includes a tiny camera that plots the precise position of the pen on a special type of paper, which the LogiTech io2 software converts into digital documents. The software also has handwriting recognition capability, so written words can be converted into text. The pen contains one megabyte of storage and has enough battery power for a user to write up to 40 pages of notes. The data is stored in flash memory in the pen, so it is quite difficult to lose.

A key part of this technology is the paper on which the pen must be used. The paper contains a pattern of very small dots that are spaced .01 inches apart. Logitech sells a wide variety of paper types, including Post-it Notes, spiral-bound notebooks, and notepads. Expect to pay about 10 cents per page.

The accountant mostly works with very specific forms, not blank pads of paper. A South Dakota-based company, Talario, has invented a variation on Logitech's system whereby any forms a company needs can be printed onto the special paper required to use the digital pens. In brief, load the special paper into a printer, print the forms using their existing application software, and write on the paper with a digital pen. Then upload the data on the pen, and the Talario software generates a PDF document that includes both the original form and the writing from the pen. This system also works in a network environment, so that forms can be printed on one printer, and then written upon with multiple pens whose contents are then uploaded through other computers in the network.

Examples of possible forms to use with a digital pen are audit checklists, contracts, purchase orders, credit applications, signature cards, W-4 or I-9 forms, employment applications, time reports, and sales orders.

In short, the digital pen completely eliminates the need to scan or fax paper documents, instead creating digital documents that can be emailed elsewhere or stored within an accounting system.

KEYSTROKE ERROR REDUCTION

One of the key problems that managers tend to overlook is the massive disparity in the error rates for the keying of alphanumeric data versus numeric data. There are many studies on this topic whose results vary by such factors as the size of the keyboard used (bigger is better) and the skill of the typist. However, the relationship between the two error rates is clear enough—entering alphanumeric data through a keyboard is 100 times more likely to result in an error than numeric data entered through a keypad, assuming the skills of a normal 60 word-per-minute typist. The error disparity drops to 25x for an expert typist, but the difference is still enormous.

How does this impact the accountant? Clearly, it makes a great deal of sense to replace as much alphanumeric data as possible with numeric data. For example, warehouse bin locations could be converted to numeric codes instead of the more common Aisle A, Rack 3, Bin R style. Similarly, product codes should be purely numeric. Also, don't try to create different versions of an invoice, such as invoice number 1234A. Further, avoid using transaction identifiers, such as a purchase order that is coded as PO-45678. By focusing hard on using just numeric codes, an accounting department can drastically reduce its keypunching error rates.

ELECTRONIC DOCUMENT STORAGE SYSTEMS

The traditional accounting data storage system is the filing cabinet—and lots of them. As soon as a transaction is completed, it is printed out and filed. However, if the accounting department is a large one, the volume of documents can be exceptionally large, which causes clutter, misused space, lost documents, and requires a large team of clerks to file and retrieve documents. A better approach is an electronic document storage and retrieval system.

This system begins with a clerk who scans all accounting documents, such as supplier invoices, into a digital format, and assigns several index codes to them so they can be searched for with several different key words. The digitized document is usually stored on an optical disk array, which has a great deal of storage capacity. Since this form of storage tends to have a rather slow data retrieval speed, the indexes are instead kept on a higher-speed computer.

Whenever a user wants to retrieve a document, he can type a search word (such as the invoice number, the name of a supplier or even a dollar total) into a computer terminal and see the document image appear in a few moments, after the index key has located the image in the optical storage device. This approach has several advantages. First, as long as multiple indexes are used, it is very difficult to ever lose a document image in the database (as opposed to a paper-based system, where documents are readily mis-filed). Second, there is only a minimal amount of storage space required near the accounting department— the original documents can be discarded unless there is a legal reason for keeping them, or can at least be sent off to less-expensive warehouse storage. Third, employees can call up the same document image at the same time, instead of having to wait in line for the same documents. Fourth, the document search time is completely eliminated, which greatly increases the efficiency of the accounting clerical staff. These represent serious efficiency and effectiveness improvements, and make this a solution well worth considering.

The main failing of the electronic document storage system is its cost. A very small installation with adequate in-house computer support may cost as little as $25,000, but will have primitive functionality and limited storage capacity. A more likely minimum cost expectation for a small company is an expenditure of $100,000, with really large accounting shops requiring much heftier storage and scanning capabilities that require an investment of more than $1 million. Nonetheless, the operational improvements resulting from such a system are so significant that it is at least worthwhile to conduct a cost-benefit analysis to see if there is a justifiable basis for investing in such a system.

DATA WAREHOUSE

Even the most perfectly maintained general ledger will not yield a sufficient amount of management information. One problem is that there may be several company locations, each maintaining its own general ledger; if there is a need for consolidated information that summarizes the contents of all these general ledgers, the only option is to manually combine the required information, which is both time-consuming and subject to error. Another problem is that many general ledgers only allow for the storage of information for a limited amount of time, such as the last two years. All earlier information must be deleted, which makes it difficult to create multi-year comparison reports. Yet another issue is that the general ledger usually does not allow for the storage of non-financial information, even though various types of operating data are at least as important as financial information. As a result, other types of information are stored in alternative databases throughout a company, and can only be accessed or consolidated with difficulty. All of these problems can be eliminated through the use of a data warehouse.

A data warehouse is a central storage facility for a company's most crucial information. This may include accounting, marketing, production, or other types of data. It can contain information that is collected over long intervals, such as once a month, but is much more effective if it contains up-to-the-minute information, so that reports reveal the most current status of a company's condition. It can also contain data for any time period that a company desires—be it one year or the past decade, or longer intervals for some types of data than for others. It can also contain data from all possible company locations. Though this may sound like a massive repository, that is not the case; the data warehouse is designed to issue information for further use in a variety of applications, which may allow the data warehouse designers to screen out data that has not further use than at the local level, where it was generated. Thus, the data warehouse is really a subset of the most useful data that are collected throughout a company, concentrated in one central database.

What are the uses of a data warehouse? Some of the more common ones are:

- *Real-time reporting.* Anyone with access to a data warehouse can use it to access up-to-the-minute data about activities throughout the company, as opposed to waiting for the more formal reports issued by the accounting department at the end of each month. However, such rapid data accessibility has its price, as noted later in this section.

- *Drill-down capability.* With so much data stored in the data warehouse, one can easily trace summary-level information back through the supporting levels of data, so that a user can easily find the supporting details without having to go to the accounting department to request that a formal search for the information be made. This greatly reduces the research work of that department, though it does require some training of data warehouse users to ensure that they can use the drill-down feature.

- *Exports to standard on-line reports.* The data warehouse can be used as a source of data for the construction of a standard set of reports that are automatically posted on-line for general company access (also known as an executive information system, or dashboard). This type of report may be updated in real time or somewhat less frequently. The main point is that the entire process is completely automated, so the accounting staff is not called upon to create any reports.

- *Automated tax filings.* If all sales tax-related information is shifted into the data warehouse, as well as the format for each tax report that is used to report to each state, then the entire process of creating tax returns can be partially or fully automated. This vastly reduces the work of the taxation staff, which can shift its efforts from the highly laborious chore of filling out tax forms to the much more effective task of developing tax strategies that will reduce a company's overall tax burden.

Though these advantages of a data warehouse sound ideal to any company that has trouble marshalling its data, there are a number of issues that make this a difficult concept to implement, including:

- *Cost.* Because a data warehouse requires a great deal of custom programming to create a separate database plus a set of automated reports, as well as interfaces to various files throughout an organization that pull in data for central storage, the cost of a data warehouse can be enormous. Any installation costing less than $1 million should be considered a triumph, with project costs exceeding $10 million being the norm for larger corporations. Also, the data warehouse requires constant updating, due to the need to include and exclude data as a company constantly modifies its requirements for various kinds of information. This will require the services of a number of full-time programmers.

- *Creation and modification of standard reports.* Though the output generated from a data warehouse can be highly automated, this does not mean that users will not want changes to the output from time to time, as their data requirements change. Whenever this happens, a considerable effort may be required to find the requested data, create an interface to the data warehouse, modify the database structure, and then include it in an automated report.

- *Data accuracy.* One of the principal reasons why accounting managers like to have complete control over any reports that contain financial information is that they want to review the data for accuracy before they issue them to the company. This error-checking step goes away when a data warehouse is used, since users can now access the data directly. To ensure that data accuracy continues to be at a high level requires the installation of any number of control systems at the user level, as well as the assistance of the internal audit and systems analysis personnel to ensure that accuracy problems are spotted and corrected at their sources. This is a prolonged and sometimes quite expensive task.

- *Data collection.* When a data warehouse is created, an interface must be built between every data source and the data warehouse. For a larger corporation, this may require the creation of hundreds of interfaces that are linked to databases scattered throughout an organization. This can be a huge task, greatly influencing the cost and duration of the installation of the data warehouse.

- *Database customization.* The database of the data warehouse must be entirely customized, since there is not a packaged database that will meet every possible data need. The configuration of the database is an extremely detailed and expensive process that can take many months to complete, especially if there is a large quantity of different types of data to be stored.

- *Frequency of data updates.* The cost of refreshing data in the warehouse can be high if the frequency of the updates increases. For example, if data were to be extracted from a subsidiary-level general ledger every few minutes, this would require the storage of every data update. Consequently, it is more common for a data warehouse to refresh its data no more frequently than once a day.

- *Long-term data storage.* Though it is possible to store an infinite quantity of data, going back through many years of a company's history, the cost of such data storage may limit a company to the storage of only a few years of data, or to the long-term storage of only a few types of data. The length of data searches can also become excessive if the database becomes too large.

Given the significant difficulties involved with the creation and ongoing maintenance of a data warehouse, it is no surprise that it is only installed by companies with hefty IT budgets and a tolerance for long-term development time frames.

SUMMARY

A number of new technologies can greatly reduce the amount of effort that the accounting department has traditionally put into data collection and storage. Some are well-known and accepted in the marketplace, such as bar coding and computerized time clocks. Others are quite new, such as digital pens, and have yet to achieve a significant level of market penetration. An accounting manager should not feel compelled to use all of these technologies. Instead, adopt only those that can achieve a discernible level of cost reduction, or assist in meeting a specific strategic or tactical goal. Another valuable use of these systems is to shift the accounting department's focus away from the manual collection and storage of data, and into more value-added analysis and reporting activities that will result in a greater level of service to the entire organization.

Chapter 11

Process Documentation

An important tool for achieving revised accounting systems is process documentation. It is necessary to thoroughly document the flow of a process before it can be streamlined. After streamlining has occurred, the new process must be carefully documented so that the people performing the function can understand how to do it. Without proper understanding, mistakes will occur in the process, leading to considerable costs to correct the mistakes.

This chapter explains how to document both existing and prospective processes, using flowcharts and written procedural documentation that are combined into a procedures manual. The procedure being documented in this chapter as an example is the order-taking process.

DOCUMENTING EXISTING PROCESSES

Documenting an existing process is not as simple as listing the sequence of events in the process. If we were reviewing an order-taking process in such a manner, we might derive the flowchart shown in Exhibit 11.1. Note the minimal number of flowchart symbols needed. An oblong shows a process, and a diamond shows a decision point.

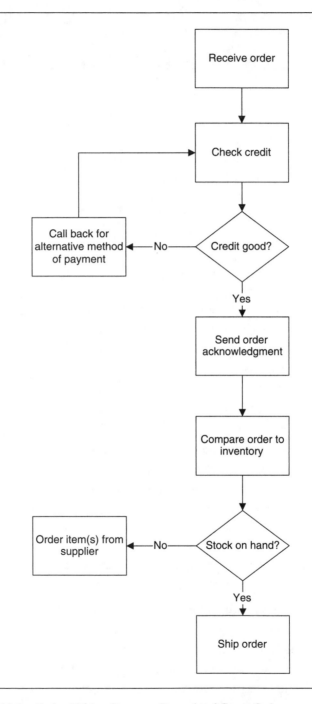

Exhibit 11.1   Order-Taking Process: Procedural Steps Only

Many other flowchart signs exist, but these two signs can be used to document many processes.

This initial flowchart shows the processing steps but leaves out several key items:

- *Time required to complete each step.* Processing times must be documented, so that process improvement work will be directed toward shrinking the processes that take a long time, rather than the short-time processes.

- *Wait time between processing steps.* Usually longer than the actual processing time, the wait times must be documented to highlight areas requiring improvement.

- *Forms used as inputs.* Forms need to be documented in order to highlight those with duplicate information, or unused or missing information.

- *Reports used as outputs.* Frequently, more reports are being generated than anyone suspects, because reports were requested for short-term needs and never cancelled once the needs were fulfilled. Also, the information on several reports may be merged onto fewer reports.

- *Who performs each step?* Wait times are often so long because the process steps jump between employees, and the wait time occurs because the item being processed enters each person's work queue and waits there while previous work-in-process is completed. A typical remedy is to concentrate consecutive task steps with the smallest number of employees, thereby reducing the number of work queues.

- *Information communication paths.* When a person needs information that is not being supplied by the formal system, the employee will go outside the system to locate the necessary information. By documenting these informal communication paths, they can be added into the new formal process.

- *Key control points.* Control points must be fully documented, so that changes can be made to the system with full knowledge of which controls will be affected by the changes.

- *Errors.* It is important to find out where errors occur in the process, since error correction is very time-consuming. To do this, the average number of errors per thousand can be listed for each process step. An internal audit study of the process will be required to collect this error information.

Exhibit 11.2 shows the same flowchart, but includes process and wait times. This information allows the user to prioritize activities that should be revised in the new process. It is clear from the exhibit that wait times are the primary culprit in causing long processing intervals. In particular, the wait time before comparing the order to inventory and the wait time before shipping the order keep a typical order from being processed in just a few hours.

There is little reason to spend time and money in enhancing the Receive Order step, since it only consumes 10 minutes. However, other considerations may require improvement of otherwise acceptable steps. For example, the marketing department may want to use "caller ID" in order to call up existing customer records prior to answering a phone call. This need would only slightly speed up the process but could be entered in the accompanying documentation of prospective changes and incorporated into the finished process.

Exhibit 11.3 shows the same flowchart, but includes the forms used as input and the reports used as output. It shows an order form, which could be replaced by online entry directly into the computer, and a cumbersome process of creating individual purchase orders for each item requiring restocking, when in-quantity ordering may be simpler.

Exhibit 11.4 shows the same flowchart but includes the position or department performing each step. It shows when the process switches between people and enters the work queue of a new person. The process has a number of hand-offs between employees, resulting in long wait times between process steps. Based on this information, consolidating process steps with fewer people should eliminate a portion of the wait time.

Exhibit 11.5 shows the informal communication paths taken by employees to perform their assigned tasks. The amount of information

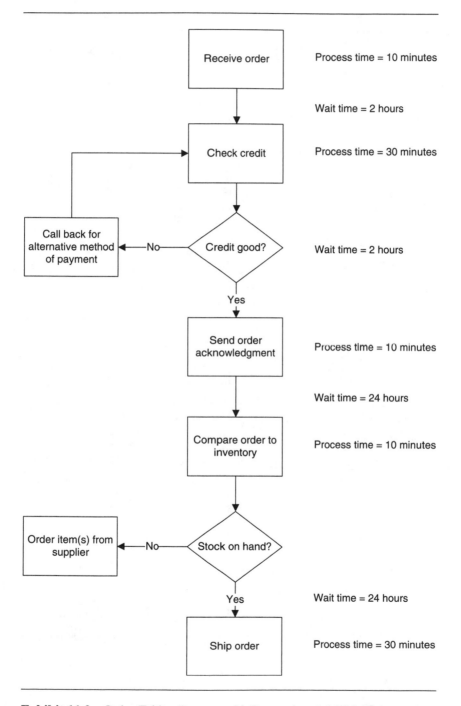

Exhibit 11.2 Order-Taking Process with Processing and Wait Times

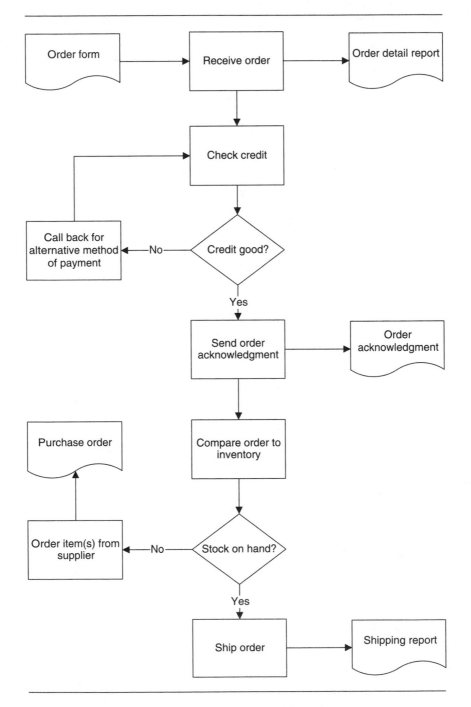

Exhibit 11.3   Order-Taking Process with Forms and Reports

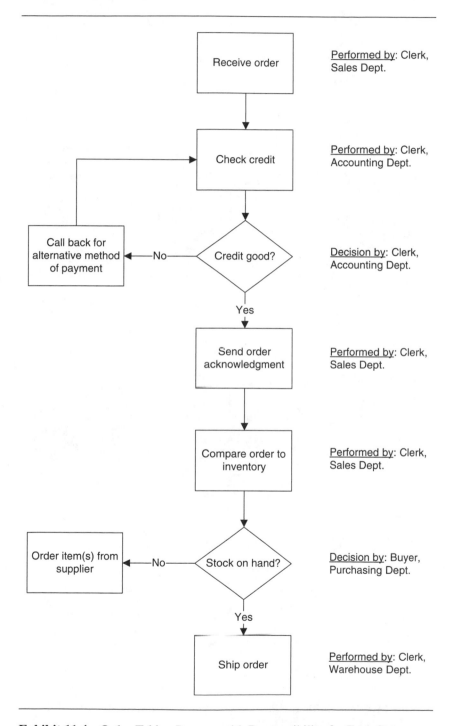

Exhibit 11.4 Order-Taking Process with Responsibility for Each Step

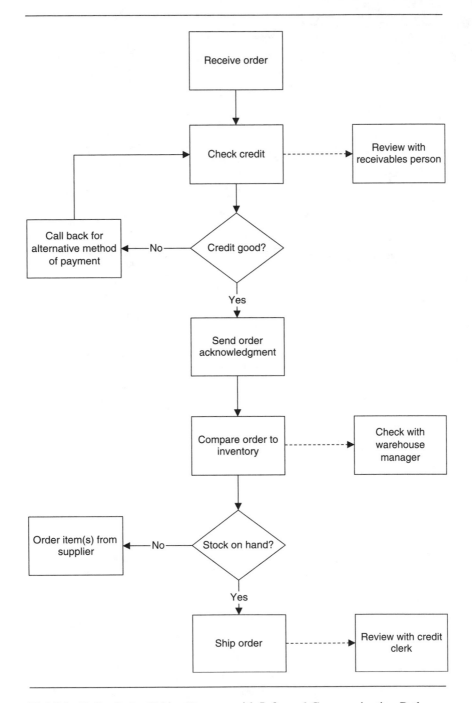

Exhibit 11.5   Order-Taking Process with Informal Communication Paths

communication hints at inaccurate information in the corporate database. Since information communications are being used to verify computer information in the areas of credit and inventory, it appears that the employees do not trust the information they are receiving from the system, and are verifying it through informal networks. Obviously, audits of receivables and inventory accuracy are needed to find the underlying problems causing the inaccurate information.

Exhibit 11.6 shows only those interdepartmental controls related to not losing orders in the system. The presence of so many cross-checks indicates that loss of orders has been a problem in the past.

Exhibit 11.7 shows the average number of errors per thousand for each step of the order-taking process as well as a breakdown of the specific error types. The data indicates a considerable problem with entering accurate information into the beginning of the process, with the effect cascading through the entire process. For example, errors are occurring in entering accurate customer addresses during the order-taking step, resulting in products being shipped to incorrect addresses at the end of the process. This problem has many solutions, such as on-line editing as orders are entered into the computer (e.g., do zip codes match the entered city name?), reading back item descriptions to customers over the phone, and auditing inventory accuracy levels, with follow-up on recommended inventory-tracking suggestions.

DOCUMENTING PROSPECTIVE PROCESSES: FLOWCHART UPDATES

After the existing process has been reviewed, the company decides to make the following changes to the order-taking process:

- Merge the order-taking and credit review staffs, so that one person can handle both tasks. This will eliminate the 2-hour wait time while transactions flow between departments.
- Give the order-taking/credit-review clerk read-only access to on-line inventory records, thereby eliminating the 24-hour wait time

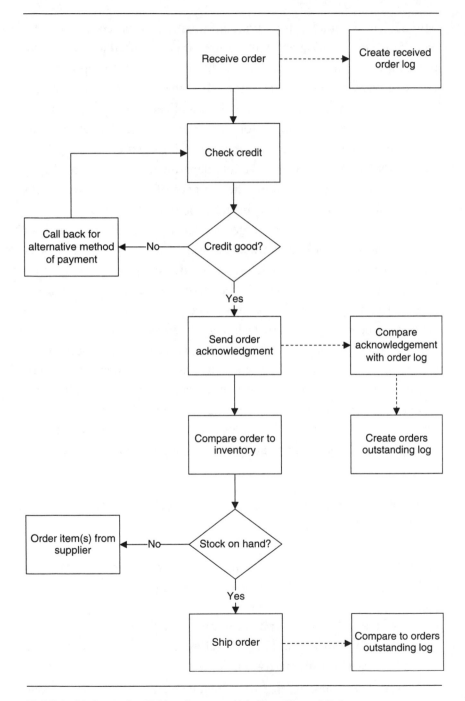

Exhibit 11.6 Order-Taking Process with Key Control Points

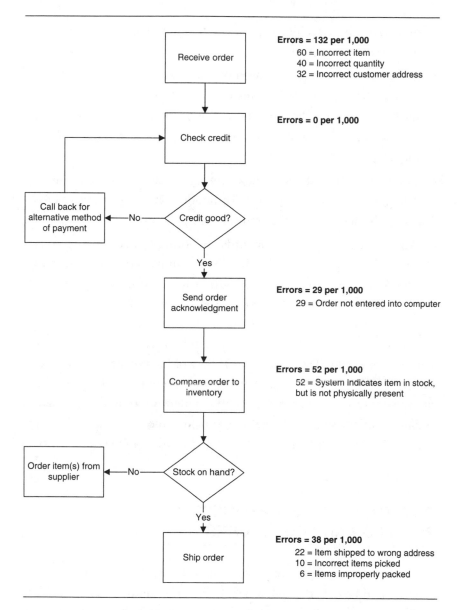

Exhibit 11.7   Order-Taking Process with Errors Noted

for information to come from the warehouse (this improvement assumes a computerized perpetual inventory system).

- Automatically create a purchase order for items not in stock, to be reviewed by a buyer before transmission to a supplier. This eliminates the possibility of out-of-stock items not being reordered and speeds paper flow.

- Initiate periodic audits of inventory accuracy, leading to higher accuracy levels. This will eliminate the need for information communications to cross-check the accuracy of the inventory records and will decrease the number of errors in the Compare Order to Inventory step.

- Initiate periodic audits of the accuracy of customer credit history. This eliminates the need for information communications to cross-check the accuracy of the credit records.

- Automatically compare records of orders received to orders shipped, and periodically audit the remaining orders to ensure that no orders are being lost. This will reduce the need for continual cross-checks at every step of the order-taking process.

The flowcharts shown earlier can then be revised to show the new sequence of events. The basic process flow of the prospective process is shown in Exhibit 11.8. Once the user has a summary flowchart that shows how to process an order, more detailed information is added to amplify upon the summary-level steps noted in the flowchart. This information is written into a procedure, which is covered in the next section.

DOCUMENTING NEW PROCESSES: POLICIES AND PROCEDURES

Procedure manuals (despite their name) actually contain policies, procedures, and flowcharts. Job descriptions are sometimes added to clarify the roles of certain positions. A *policy* is a statement by top management to be used as a general guide by the organization on how to deal with a situation. A *procedure* is a specific rule or series of steps to follow in accomplishing a task. Procedures can be developed at any level in an organization. It is when policies and procedures are written down that manuals come into being.

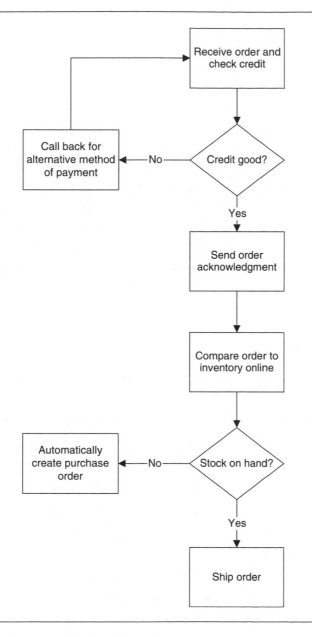

Exhibit 11.8   Revised Order-Taking-Process

All entities have policies and procedures governing their various transactions. However, many have only informal systems whereby policies and procedures are created and modified; these systems result in inconsistent treatment of similar activities with resulting inefficiency and loss of uniformity. It is therefore important that the policies and procedures be written down and published.

The procedure manual is an instrument of communication. It provides a common understanding of policy interpretations and clearly states the steps to be performed to complete tasks.

As a business expands, responsibility and authority are delegated to a greater extent. Manuals then become more important, since they are a key tool by which management communicates its goals, and controls the implementation of those goals. Also, manuals bind together dispersed but similar operations. They are the device by which a company says, "This is how we do it." Companies may create manuals to fulfill any of the following needs:

- To communicate to all subsidiaries the key policies and procedures of the company.
- To provide a common understanding of policy interpretations and to define and clarify policy or procedural issues that may arise.
- To promote standardization and simplification.
- To train new employees, or those assigned to new jobs.
- To rapidly implement new procedures throughout the company.
- To allow timely communication of changes in policies and procedures.
- To create an internal control tool for management and form the basis for an audit of performance for compliance.
- To avoid duplication of effort by clearly stating responsibilities. A manual usually includes complete job descriptions for key positions.
- To reduce the amount of management time required to provide instructions and directions.

As noted earlier, a policy is a statement by top management to be used as a general guide by the organization on how to deal with a situation. Exhibit 11.9 lists several policies pertaining to the order-taking example. The policies in the exhibit clearly state the boundaries within which personnel may operate. For example, requested credit of $50,001 to $100,000 must be approved by the credit supervisor. An explicit statement like this is less subject to misinterpretation than "excessive credit must be reviewed by management." Clearly stated policies lead to minimal staff confusion, speedier transaction processing, and minimal errors to correct.

As noted earlier, a procedure is a specific rule or series of steps to follow in accomplishing a task. Exhibits 11.10 and 11.11 list several procedures pertaining to the order-taking example. All are very specific. Without clear instructions, an employee will have to guess at the correct sequence of events and will likely create errors by doing so.

Exhibit 11.9   Order-Taking and Related Policies

Policy: Credit Limits   **Page 1 of 1**

Any orders received that are not prepaid and exceed $50,000 will be
forwarded to the credit supervisor for credit approval. Any such orders that
exceed $100,000 will be forwarded to the Treasurer for additional approval.
Any such orders that exceed $250,000 will be forwarded to the Chief
Financial Officer for additional approval.

Policy: Order Filling   **Page 1 of 1**

All orders with stock on hand will be filled within two days of order receipt.
If stock must be back-ordered, the order will be shipped with all
possible items within two days of order receipt; back-ordered items will
be shipped within one day of receipt.

Policy: Inventory Audits   **Page 1 of 1**

All on-site raw materials will be audited once a week. The audit team will
review the quantity, unit of measure, and location data for a statistically
significant proportion of the inventory. All off-site inventory will be reviewed
monthly, using the same criteria.

Exhibit 11.10   Credit Review Procedure

Procedure: Credit Review   **Page 1 of 1**

1. Access the Dun & Bradstreet website.
2. Enter the phone number of the company under review, which brings up the credit record.
3. Purchase a credit report and have it emailed to the company.
4. If average payment days on the credit report are below 40, accept credit up to a limit of $50,000.
5. If average payment days are over 40, review payment history on previous receivables. If average payment days are below 40, accept credit up to a limit of $50,000.
6. If there is no previous receivable within the past two years for the company under review, give the file concerning the company to the credit supervisor for further review.

Consequently, it is better to write procedures with too *much* detail than with too *little*.

An accounting manual tells employees how to do their jobs. However, policies and procedures alone do not specifically state who is to complete a task. Descriptions aggregate procedures by specific job, and so are a useful part of a policies and procedures manual. A job description should list the following items:

Exhibit 11.11   Order Acknowledgment Procedure

Procedure: Send Order Acknowledgment   **Page 1 of 1**

1. Access the order entry software and enter the Review Inventory Status screen.
2. Enter the part numbers and quantities from the customer order.
3. The software reserves inventory for the order.
4. If necessary, enter Y for items that must be placed on back-order.
5. Once the order has been completely entered, enter Y at the Print Order Acknowledgement prompt.
6. Accept the delivery option of emailing the acknowledgment to the customer.

- *Reports to.* Employees may not know who they report to. Identifying the responsible party eliminates all doubt about who is authorized to assign tasks to an employee.

- *Supervises.* This clearly identifies staff reporting to the position, and clears up any confusion with other managers regarding this issue.

- *Basic function.* Briefly states the general tasks that the position is responsible for handling. This is useful in case the more detailed accountabilities listed in the following section mistakenly do not include an item that clearly falls within the general range of the job.

- *Principal accountabilities.* The procedure name or reference number can be listed next to each responsibility, which an employee can use to reference more detailed information about each task. Reports generated should also be listed, with a reference number that links it to an example document.

- *Update date.* Tasks, staff responsibilities, and reports change constantly, so the job description must be updated frequently to reflect the alterations.

Exhibit 11.12 is an example of a chief financial officer's job description.

Exhibit 11.12   Chief Financial Officer Job Description

Position Name: Chief Financial Officer
Reports to: Chief Executive Officer
Supervises: Controller, Tax Manager, Human Resources Manager, Investor Relations Officer

Basic Function: This position is accountable for the administrative, financial, and risk management operations of the company, to include the development of a financial strategy, metrics tied to that strategy, and the ongoing development and monitoring of control systems designed to preserve company assets and report accurate financial results.

(Continued)

Principal Accountabilities:

1. Understand and mitigate key elements of the company's risk profile

2. Construct and monitor reliable control systems

3. Develop financial and tax strategies

4. Develop performance measures that support the company's strategic direction

5. Manage the treasury, accounting, investor relations, tax and human resources departments

6. Monitor financial reports

7. Manage the capital request and budgeting processes

8. Oversee the issuance of financial information

9. Implement operational best practices

10. Supervise acquisition due diligence and negotiate acquisitions

11. Maintain banking relationships

12. Arrange for equity and debt financing

13. Invest funds

14. Invest pension funds

15. Maintain appropriate insurance coverage

Update date: July 1, 2009

SUMMARY

Flowcharting and related systems documentation are needed to identify the parts of existing processes that can be streamlined and should also be used to lock in improved processes, so that employees will know the correct methods for completing them.

Procedure manuals are needed to tell personnel how to complete procedures in the correct manner, thereby avoiding errors that must later be corrected. Each manual should include a clear flowchart of each process, written descriptions of each step, related policies, and those job descriptions that will be responsible for the targeted procedures.

While useful, a procedures manual will rapidly become outdated unless continually matched against existing practices. Consequently, a staff person should be assigned the task of policy and procedure monitoring and updating.

Chapter 12

Change Management

This chapter discusses the effect on employees of the many changes advocated in this book. Far more than a hundred changes are listed in the previous chapters, and implementing even a fraction of them will cause significant disruption in the work of the accounting department. One study shows that 84% of the labor in the accounting and finance departments is devoted to transaction processing—precisely the area that this book focuses on shrinking, meaning that the majority of accounting employees will be affected by these changes. This chapter warns the reader of organizational problems caused by changes, details the effects of disruption in the organization, notes when change is most likely to succeed, and explains how to deal with it in a positive manner.

EFFECTS OF CHANGE ON THE ORGANIZATION

Many improvement projects never reach completion because the organization rejects them. Minor changes have a greater chance of not being rejected by the organization, since employees do not feel threatened by them, but major changes require a quite different implementation approach.

A major change probably will not be accepted by an organization for a variety of reasons. First, many employees are comfortable operating within the set of procedures that define their jobs. By changing the procedures, their jobs are changed, which causes them considerable uncertainty. They may react by trying to reintroduce their comfortable former procedures. Second, many changes will cost employees their jobs; reaction to such changes will be understandably negative. Also, changes may require new lines of authority, so that employees find themselves reporting to new supervisors. Finally, certain types of employees react negatively to all types of change and will continue to do so no matter how well they are educated about the reasons for change. Any or all of these problems will surface during a major implementation.

When a major change is being implemented, an organization will pass through a well-defined set of responses. The management team will initially be optimistic about the overall objective, but will then turn more pessimistic when they determine the extent and cost of change. Many projects quietly vanish at this point. If the project proceeds, it will be rewarded with a cautious sense of optimism, especially if it stays close to the original time and cost budget. There will then be a period of rejection at the end of the project when the new system replaces the old one, and questions are raised (perceived or real) with the new system. It is still possible for a project to fail at this point, at the cusp of success. Finally, after the last objections are overcome, the organization grudgingly accepts the new system.

When an organization makes lightning-fast changes, it becomes so unstable that the work force becomes transient. When whole departments are formed and dissolved over very short time periods, it becomes evident to the employees that the company is unwilling to provide any job tenure, resulting in increased turnover.

In summary, a project must pass through several phases when an organization is likely to reject it, and changes may result in major staff turnover. How does the management team ensure that projects will survive this obstacle course of emotional trauma and reach a successful conclusion? The next section offers some suggestions.

HOW TO OVERCOME ORGANIZATIONAL RESISTANCE TO CHANGE

Project implementation teams tend to focus on the technical aspects of their projects, and are taken aback when the organization fails to support their projects for reasons entirely unrelated to the resulting benefits. However, by being aware of the various reasons why organizations reject projects, it is possible to circumvent the problems presented by change.

The first step the management team should take is to ensure that a project has a strong sponsor. This must be a senior-level manager who is deeply committed to the project, and who is willing to intercede for the project team whenever necessary to ensure that the project will be a success. When the organization realizes that this senior manager will not back down from the project, it will realize that the project is inevitable, and will be much more likely to support it themselves. However, this can be construed as forcing a project onto an organization. In addition to having a sponsor, it also makes sense to persuade the organization itself to accept the changes. How can this be done?

The project team must meet with a variety of employees who will be affected by their new project, and show them the problems that will occur if the project is not completed on time, such as higher costs that lead to future layoffs or longer processing times that make the company less competitive and that also may lead to layoffs. The delineated crisis must be real, or else employees will suspect that management is making up reasons for forcing the project upon the company. This gloomy view of the future can be followed by a clear picture of the benefits to be gained from the new system, as well as a detailed, step-by-step plan for how to get to that point. This information should be laid out in as much detail as possible, so that employees know that the project team (and management) is not hiding any information from them. As the project progresses, team members should meet with employees periodically, and update them on the project's progress. This sharing of information helps to gain the acceptance of the organization for the project.

If information sessions are not sufficient, the project team should also schedule education sessions for key users from the organization that gives them an in-depth view of the changes being implemented. Training can come from many sources. The information contained in training classes can come from experimenting with new ideas, comparisons with other organizations, and formal training classes. This level of involvement is sufficient to persuade the majority of employees that the project should be implemented.

There will always be a few employees who are unchangeably comfortable with the status quo, and who will attack any proposed changes that will alter the current set of procedures that define their jobs. The project team should listen to the concerns of these people and address any problems that can be accommodated without jeopardizing the successful (and timely) completion of the project. However, if they continue to oppose the project after all reasonable efforts have been made to accommodate them, they should be removed from that part of the company that is affected by the project. Though a painful step to take, it is necessary to avoid their continued sniping at the project, possibly even after the project has been implemented.

If some employees are shifted out of the project area as a result of a new project, new ones must be hired in. This is a considerable opportunity for management to hire the right kind of employees who readily accept change. Interviews for these new positions should be rigorous, focusing in particular on candidates' level of comfort in a changing environment, current technical skills, and willingness to assist further change efforts.

In short, if the project team has a strong sponsor, communicates a detailed project plan to the organization, removes "snipers" from the focus area, and hires new employees who are comfortable with change, then the project team has an enhanced chance of success. The general environment is also a strong determinant of success, as discussed in the next section.

THE MOST FERTILE GROUND FOR CHANGE

Before making any type of systems improvement, it is useful to review the existing environment to see if there is a reasonable chance for it to succeed. The following bullet points note the best environments in which a best practice can not only be installed, but also have a fair chance of continuing to succeed:

- *If benchmarking shows a problem.* Some organizations regularly compare their performance levels against those of other companies, especially those with a reputation for having extremely high levels of performance. If there is a significant difference in the perform-ance levels of these other organizations and the company doing the benchmarking, this can serve as a reminder that continuous change is necessary in order to survive. If management sees and heeds this warning, the environment in which a project will be accepted is greatly improved.

- *If management has a change orientation.* Some managers have a seemingly genetic disposition toward change. If a department has such a person in charge, there will certainly be a drive toward many changes. If anything, this type of person can go too far, implementing too many projects with not enough preparation, re-sulting in a confused operations group whose newly revised sys-tems may take a considerable amount of time to untangle. The presence of a detail-oriented second-in-command is very helpful for preserving order and channeling the energies of such a manager into the most productive directions.

- *If the company is experiencing poor financial results.* If there is a significant loss, or a trend in that direction, this serves as a wake-up call to management, which in turn results in the cre-ation of a multitude of improvement projects. In this case, the situation may even go too far, with so many improvement proj-ects going on at once that there are not enough resources to go around, resulting in the ultimate completion of few, if any, of the projects.

- *If there is new management.* Most people who are newly installed as managers want to make changes in order to leave their marks on the organization. This can include a renewed focus on efficiency that will result in the implementation of multiple best practices.

In short, as long as there is willingness by management to change and a good reason for doing so, then there is fertile ground for the implementation of a multitude of improvements. But even if the general environment is conducive to change, the timing of that change is also crucial to success, as described next.

THE TIMING OF CHANGE

The timing of a project implementation, the time it takes to complete it, and the pacing of installations have a major impact on the likelihood of success.

The timing of an implementation project is critical. For example, an installation that comes at the same time as a major deliverable in another area will receive scant attention from the person who is most responsible for using the best practice, since it takes a distant second place to the deliverable. Also, any project that comes on the heels of a disastrous implementation will not be expected to succeed, though this problem can be overcome by targeting a quick and easy project that results in a rapid success—and which overcomes the stigma of the earlier failure. Further, proper implementation timing must take into account other project implementations going on elsewhere in the organization or even in the same department, so that there is no conflict over project resources. Only by carefully considering these issues prior to scheduling a project will an improvement project not be impacted by timing issues.

In addition to timing, the *time* required to complete a project is of major importance. A quick project brings with it the aura of success, a reputation for completion, and a much better chance of being allowed to take on a more difficult and expensive project. Alternatively, a project that impacts lots of departments or people, or which involves the liberal application of cutting-edge technology, runs a major risk of running for a long time; and the longer the project, the greater the risks that something will go wrong, objections will arise, or that funding will run out. Thus, close attention to project duration will increase the odds of success.

Also, the concept of pacing is important. This means that an implementation will be more likely to succeed if only a certain number of implementations are scheduled for a specific area. For example, if corporate management wants to install several dozen different types of best practices in five different departments, the best implementation approach is to install one project in a single department and then move on to a different department. By doing so, the staff of each department has a chance to assimilate a single best practice, which involves staff training, adjustments to policies and procedures, and modifications of work schedules. Otherwise, if they are bombarded with multiple projects at the same time or one after another, there is more likelihood that all of the projects will fail or at least not achieve high levels of performance for some time. In addition, the staff may rebel at the constant stream of changes and refuse to cooperate with further implementations.

DUPLICATING CHANGES

It can be a particularly difficult challenge to duplicate a successfully completed project when opening a new company facility, especially if expansion is contemplated in many locations over a short time period. The difficulty with duplicating change is that employees in the new locations are typically given a brief overview of a best practice and told to "go do it." Under this scenario, they only have a sketchy idea of what they are supposed to do, and so create a process that varies in some key details from the baseline situation. To make matters worse, managers at the new location may feel that they can create a better best practice from the start, and so create something that differs in key respects from the baseline. For both reasons, the failure incidence of change duplication is high.

To avoid these problems, a company should first be certain that it has accumulated all possible knowledge about a functioning best practice—the forms, policies, procedures, equipment, and special knowledge required to make it work properly—and then transfer this information into a concise document that can be shared with new locations. Second, a roving team of expert users must be commissioned to visit all new company locations and personally install the new systems, thereby ensuring that the proper level of experience with a best practice is brought to bear on a duplication activity. Finally, a company should transfer the practitioners of best practices to new locations on a semi-permanent basis to ensure that the necessary knowledge required to make a best practice effective over the long term remains on site. By taking these steps, a company can increase its odds of spreading change throughout all of its locations.

A special issue is the tendency of a new company location to attempt to enhance a copied best practice at the earliest opportunity. This tendency frequently arises from the belief that one can always improve upon something that was created elsewhere. However, these changes may negatively impact other parts of the company's systems, resulting in an overall reduction in performance. Consequently, it is better to insist that new locations duplicate a best practice in all respects and use it to match the performance levels of the baseline location before they are allowed to make any changes to it. By doing so, the new location must take the time to fully utilize the best practice and learn its intricacies before they can modify it.

THE REENGINEERING PROJECT

A re-engineering project usually involves a radical redesign of an existing process. Radical change requires considerable changes to the organization, which can cause major disruption among employees. This section discusses how to implement a project that involves a great amount of change.

Reengineering projects fail more frequently than they succeed. Part of the problem is that reengineering is not an incremental change—it involves tearing out the old system by the roots and installing an entirely new system that may not mesh very well with the existing organizational structure and processes. To make a reengineering project more successful, it is necessary to convince as many employees as possible of the need for change. Arguments can include loss of market share, spiraling costs, and a poor showing in comparison to the industry benchmark. Even a tour of a company that is world-class in regard to that process may be needed to convince skeptics.

Some of the people requiring convincing will be top management. If the management team does not throw its support behind the project, as well as assign one of its members direct responsibility for it, the project should not be undertaken. This is because top management may pull out all monetary and personnel support of the project if it thinks that other projects are more important, or that the benefits to be derived will not occur. Instead, the top managers need to be so enamored of the project that they defend it against all attacks.

Managers may be uncomfortable with the degree of change involved. They may be willing to make a few minor changes, but the wholesale dismantling and reassembly frequently encountered in a reengineering project can require considerable fortitude. One way to avoid this problem is to reengineer on a pilot basis, so that the entire company can see how the project is affecting just a small area of the company. If the pilot goes well, then acceptance of the project as a whole is much more likely. If the pilot does not proceed according to plan, then it may not have worked on a large scale anyways, and should be refined at the pilot level before being rolled out any further.

An approach that organizations find easier to accept is to come up with radical changes only at the planning stage, and to then implement the changes slowly and in segments, so that the organization has a better chance to assimilate the changes. However, this slow-change model embodies the risk of being too slow. If the original project is watered down and strung out over a long time, the project team supporting the project may become discouraged, and drift off to other projects. Also, if a company is in dire need of change, the implementation must proceed rapidly in order to keep the company from failing— the effect of the changes on the organization becomes a secondary consideration.

Another problem with reengineering is its dependence on advanced technology to add efficiency to a process. If the technology is too advanced, then it may not stand the test of actual usage on a day-to-day basis. To avoid this problem, the company should conduct pilot tests of new technology, or at least send a team to a company that has successfully implemented the technology to make inquiries about any issues it may have encountered during implementation. Another solution is to back away slightly from the most advanced technology and instead adopt technology that has been tested in the marketplace for a few years.

A reengineering project may fail if it is not staffed with an adequate number of people who are well trained and skilled in completing the project. This is particularly important for a reengineering project where advanced technology is frequently used, skills for handling recalcitrant employees are paramount, and project management skills for handling a diverse group of highly talented individuals is necessary. Project management skills are especially hard to come by. A top project manager must be able to keep the project within a time and money budget, schedule and administer project reviews, coordinate subprojects, and communicate about the project's progress with upper management. Keeping the expectations of the top management group in line with reality is particularly important, since there should be no fallout at the end of the project if the system delivered to the company does not meet expectations. At lower organizational levels, the project manager should also be involved with obtaining acceptance of the project by those individuals or departments that will be affected by the reengineering project. If there is resistance in those departments, the successful conclusion of the project will be in jeopardy.

Perhaps the most important issue for a reengineering project is that it must be planned adequately to convince the many naysayers who will arise when the project is announced. The plan must detail the project's staffing needs, time line, role-out schedule, training requirements, costs and benefits, and (especially) effects on the existing staff. It is best to be as detailed and forthright with this information as possible, both to convey an image of integrity to the staff and to provide information to employees who may offer valuable advice that can result in changes to the plan and contribute to its ultimate success.

In short, the reengineering project, which is the most difficult project of all to implement, can be completed by following the guidelines listed in the section that discussed overcoming organizational resistance to change. The main issue that sets apart a reengineering project from other projects is that *all* of the guidelines must be followed; otherwise, the organization most certainly will not accept the project.

SUMMARY

As more companies bring change to their employees, they find that many projects are not being successfully completed. This low success rate is caused by the discomfort the organization experiences when existing systems are replaced by new procedures, technologies, and reporting relationships. These problems can be overcome by gaining strong management support, planning each project in detail, communicating change plans to employees, moving dissenters away from the focus areas, and paying attention to the environment in which the contemplated change occurs. These practices improve the odds of a project's success.

Index